Joao Alexandre Lobo Marques · Simon James Fong
Editors

Computerized Systems for Diagnosis and Treatment of COVID-19

Springer

Editors
Joao Alexandre Lobo Marques
Laboratory of Applied Neurosciences
University of Saint Joseph
Macao, Macao

Simon James Fong
Faculty of Science and Technology
University of Macau
Macao, Macao

ISBN 978-3-031-30787-4 ISBN 978-3-031-30788-1 (eBook)
https://doi.org/10.1007/978-3-031-30788-1

This Springer imprint is published by the registered company Springer Nature Switzerland AG
The registered company address is: Gewerbestrasse 11, 6330 Cham, Switzerland

Contents

Technology Developments to Face the COVID-19 Pandemic: Advances, Challenges, and Trends

Joao Alexandre Lobo Marques and Simon James Fong

Abstract The global pandemic triggered by the Corona Virus Disease firstly detected in 2019 (COVID-19), entered the fourth year with many unknown aspects that need to be continuously studied by the medical and academic communities. According to the World Health Organization (WHO), until January 2023, more than 650 million cases were officially accounted (with probably much more non tested cases) with 6,656,601 deaths officially linked to the COVID-19 as plausible root cause. In this Chapter, an overview of some relevant technical aspects related to the COVID-19 pandemic is presented, divided in three parts. First, the advances are highlighted, including the development of new technologies in different areas such as medical devices, vaccines, and computerized system for medical support. Second, the focus is on relevant challenges, including the discussion on how computerized diagnostic supporting systems based on Artificial Intelligence are in fact ready to effectively help on clinical processes, from the perspective of the model proposed by NASA, Technology Readiness Levels (TRL). Finally, two trends are presented with increased necessity of computerized systems to deal with the Long Covid and the interest on Precision Medicine digital tools. Analyzing these three aspects (advances, challenges, and trends) may provide a broader understanding of the impact of the COVID-19 pandemic on the development of Computerized Diagnostic Support Systems.

J. A. Lobo Marques (✉)
Laboratory of Applied Neurosciences, University of Saint Joseph,
Estrada Marginal da Ilha Verde, 14-17, Macao SAR, China
e-mail: alexandre.lobo@usj.edu.mo

S. J. Fong
Faculty of Science and Technology, University of Macau, Macau SAR, China
e-mail: ccfong@um.edu.mo

J. A. Lobo Marques and S. J. Fong (eds.), *Computerized Systems for Diagnosis and Treatment of COVID-19*, https://doi.org/10.1007/978-3-031-30788-1_1

1 Introduction

During the last years, the global COVID-19 pandemic triggered by the Corona Virus Disease, firstly detected in 2019, has created strong and significant impacts on management practices for public health systems, definition of health indicators, and the development of new technologies. With an unprecedented speed in human history, while the most devastating disease since the global alliance after the second World War created the World Health Organization (WHO) was exponentially spreading, scientific achievements and technology advances were obtained worldwide in crucial areas to fight the pandemic, such as new vaccine development processes, low-cost medical devices (from centrifuges to mechanical ventilators), computerized diagnostic systems using data from clinical laboratory analysis, medical imaging and biosignals analysis systems.

The COVID-19 Pandemic entered through many different short-term cycles of infections, and consequently number of deaths, in just a few years in different parts of the world, which usually happened according to the restrictive public health measures and the surges of new variants of the virus in different countries. Despite the better understanding of the disease, it still presents many challenges and several unknown aspects that need to be continuously studied by the medical/health and academic communities. According to the World Health Organization (WHO), until January 2023, more than 650 million cases were officially accounted (with probably similar number of even more non confirmed cases) with 6,656,601 deaths, with a current average of 10,000 deaths per week officially linked to the COVID-19 as the most plausible root cause.

In addition to the direct and obvious consequences to the patients' health conditions, the COVID-19 created a significant burden to the public health system in every location, from developing countries to the most economically developed societies. The number of necessary resources to support the exponential demand represented a quite challenging condition, creating issues in several dimensions, such as hospital beds, intensive care units devices, mechanical ventilators, antibiotics, or even gloves, among many others. From the human side, the impact on the multidisciplinary staff working on clinical premises was enormous. The professionals were under a dual stress condition: first, the patients under extreme severe conditions and increasing number of deaths from a disease with no treatment in many cases; and, second, the personal risk of getting infected.

The harmful consequences of the COVID-19 in the whole world, created a global effort to develop efficient processes and systems to cope with the new challenges not only in the recurrent peaks, but also in a permanent way, since the disease is recurrent and a large number of patients keep different symptoms and issues for a long period of time, what is classified as Long COVID. It is important to notice that during the first twelve to eighteen months, one of the main focus was to use different computerized solutions based on epidemiological models. Since the beginning of the pandemic, even With preliminary data, classic approaches, such as the SIR/SEIR numerical models, were widely used [1]. With the pandemic global spread and more

accurate data collected, including different variants of the virus, new approaches were able to be tested [2], including nonlinear models [3] and probabilistic models based on Monte Carlo simulation [4].

Today, despite all the efforts to automatize the diagnostic through computerized systems, there is still a lack of practical applications working on clinical premises and effectively improving physicians interpretation or providing diagnostic support. The cause relies on many different aspects, such as poor integration between scientific developments and industry, access to most recent technologies from many different hospitals (especially the ones with lack of resources) and barrier to the introduction of innovative processes, creating real constraints to the adoption of new technologies. With that in mind, it becomes extremely necessary to comprehend the positive impacts for the area of healthcare of adopting computerized systems for the diagnostic and management of patients with COVID-19.

In this Chapter, a critical overview of some relevant technical aspects resulting from the COVID-19 pandemic is presented, divided in three parts. First, the advances are highlighted, including the development of new technologies in different areas such as medical devices, vaccines, and computerized system for medical support. Second, the focus is on relevant challenges, including the discussion on how computerized diagnostic supporting systems based on Artificial Intelligence are in fact ready to effectively help on clinical processes, from the perspective of the model proposed by NASA, Technology Readiness Levels (TRL). Finally, two trends are presented with increased necessity of computerized systems to deal with the Long Covid and the interest on Precision Medicine digital tools.

2 Technology and Scientific Advances

As previously mentioned, the global effort created within technical and the academic communities boosted new technology developments and scientific publications. In this section, three areas are highlighted for discussion, as presented in Fig. 1.

2.1 Development of New Vaccines

One of the major technical advances during the COVID-19 pandemic was the fast development of new technologies and the process acceleration for obtaining viable types of vaccines against the SARS-Cov-2 virus and its variants. Regulatory agencies such as the Food and Drug Administration (FDA) from the United States of America, establish new processes to speed up the approvals with the maximum possible rigor and evaluations of the clinical trials.

Many concerns were raised from different groups of health professionals and members of organized societies, in order to make it possible to verify the credibility and feasibility of the developed products. On this matter, it is important to highlight

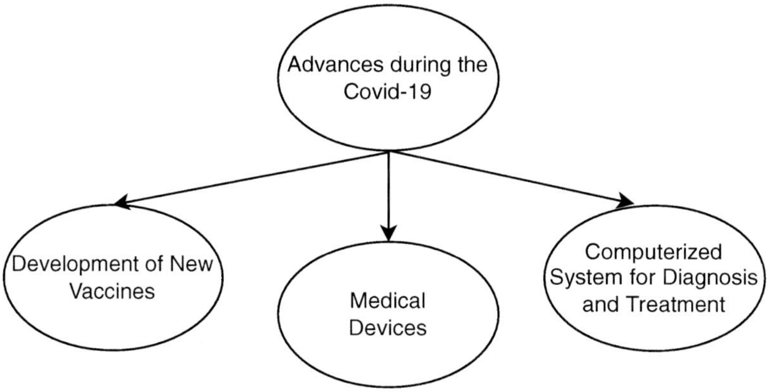

Fig. 1 Three areas of significant scientific and technological advances triggered during the COVID-19 pandemic

that only vaccines developed by important and high level companies or joint institutions were approved and released on the market, such as Pfizer, Johnson & Johnson, among a few others were considered, given the significant technical challenge.

The Mayo Clinic provides a timeline updated with simplified and non technical description, following the FDA approvals [5]. Firstly, in 2020, the FDA gave emergency approval for use to two mRNA COVID-19 vaccines, the Pfizer-BioNTech and the Moderna COVID-19 vaccines less trials, testing and retesting than is normally required, but still proving that the products are safe for human use and effective against the virus. Following, in 2021 emergency use authorization is provided to the Janssen/Johnson & Johnson vaccine. In addition, the Pfizer-BioNTech vaccine is approved for children aged between 5 to 15. More recently, in 2022, the Moderna vaccine, now called Spikevax, is authorized and the FDA authorizes the Pfizer-BioNTech for children aged between 6 months to 11 years old. Finally, the Novavax vaccine was approved for people aged more than 12 years old.

The scenario of technological development is completed with different vaccines developed in China and India, and several plants in different countries. For example, in India, currently with approximately 67% of the population fully vaccinated with at least two doses, a local production of the formula developed by Oxford and AstraZeneca is named Covishield. As another example, in Brazil, statistics indicate that 81% of the population received at least two doses of any type of vaccine, the local production of the brand Coronavac follows actually the formula of the giant Chinese pharmaceutical company called Sinovac.

Another challenge is the high number of reinfections, which occurs when after a first infection with the SARS-CoV-2 virus, the person recovers and later becomes infected again. When someone is infected for the first time, some immune protection against the virus, however, the reinfection is very common and the reasons are under study. There are ongoing studies to better understand issues related to COVID-19 reinfection, for example, how often can it occur, who is at greater risk of reinfection,

how long after the previous infection can a new infection occur, how severe/severity and the risk of transmission to third parties after re-infection. Some reasons can be stated, especially when associated, may be the cause of reinfection, such as the long duration of the pandemic, since people change the capacity of response of their immunity, making them again more vulnerable to reinfection. In addition, the vaccine immune protection provided may lose effectiveness with time, creating the challenge for public health systems to create permanent programs for administering booster dose. From the epidemiology management perspective, after stressful periods of lockdown, elevated number of deaths and economic constraints, there is a natural relaxation in some prevention measures, including protection and surveillance, creating also weak processes for data collection and decreasing the data reliability. Finally, the emergence of new variants of COVID-19, which can be more contagious, even if not as deadly as before for vaccinated individuals. For example, the omicron variant is twice as contagious as the previous ones, including delta. This variant is associated with a greater likelihood of reinfection, however, although several new variants of COVID-19 are discovered with some regularity, apparently this virus does not mutate as much as the flu virus.

2.2 New Medical Devices

The area of medical devices is vast and this Section is to discuss some relevant technology advances as a result of the exponential spread of the COVID-19 worldwide, creating significant constraints for public health systems. A key area is related to respiratory support technologies, since the use of Mechanical Ventilators with invasive ventilation has been commonly adopted for severe cases of the disease when patients are suffering from acute respiratory insufficiency (ARI). A position paper about the topic is presented in [6].

A new technology development in the area is called "ELMO" and it is a helmet for respiratory support designed and manufactured as a response to the COVID-19 pandemic. It is a non-invasive and safer respirator helmet for healthcare professionals and patients created in April 2020 by a task force involving a public-private partnership. The innovative equipment emerged as a new step for the treatment of patients with hypoxemic acute respiratory failure by Covid-19 [7, 8]. The clinical trials related to the new device are registered at the Clinical Trials portal [9]. According to the ELMO Registry survey, a research developed by the Health Research Management of the School of Public Health of Ceará (ESP/CE), an agency linked to the Health Secretariat of the State of Ceará (Sesa), 66% of the patients who used the device did not need to move to mechanical ventilator. So far, the study has evaluated about 1570 medical records of people who were hospitalized in the city of Fortaleza. Statistics indicate the rates of the need of mechanical ventilators reach 60% of the total number of patients in Intensive Care Units as a consequence of the COVID-19. With the new developed technology, this number was reduced to 34%.

2.3 Computerized Systems

The adoption of digital technologies as a result of the COVID-19 pandemic increased exponentially and the impact on clinical processes and practices was significant. This topic is the major area of the book, so an overview of how Patient's electronic data is managed, including the new pervasive use of patient's digital data, is presented. The necessity of collecting reliable data to be shared created the need of not only secure and reliable communications infrastructure but also the design and implementation of new processes.

As a major example, during the pandemic, Telehealth and Telemedicine systems were widely used as temporary solution to support patients and also to address the lack of specialists or operational staff. For example, considering Radiology, the integration of existing PACS (Picture Archiving and Communication Systems) to telecommunication systems to allow remote reporting or second opinion was already a common practice for two main reasons. One is technical. To transmit a huge number of digital images per patient to a server, generated by imaging devices such as CT-Scans, MRI's, X-Ray, Ultrasound, and to share among different medical specialists, a consolidated digital infrastructure is required. The second reason is related to resources. Radiology specialist are not easy to find and they are expensive. The possibility of setting up centralized image centers and allowing one specialist to provide support to several different institutions, even in remote locations, was a win-win situation.

With the constraints created by the COVID-19, other specializations also demanded to have support of online digital tools to have consultations, follow ups and second opinions as well. Obviously, ethical aspects were a major concern from medical associations and social institutions, to protect the patients from fraud, data leakage, or lack of privacy and confidentiality in a sensitive medical analysis. Special and temporary authorizations were granted during the period of extreme crisis and many issues related to credibility and data security were at least partially assured and the popularity significantly increased.

In Brazil, for example, as a direct consequence of the pandemic, the Federal Council of Medicine (Conselho Federal de Medicina - CFM) formally authorized and regulated the adoption of Teleconsultations [10]. Actually, the Resolution No. 2,314/2022 is broad and defines and regulates Telemedicine in the country, as a form of medical services mediated by technology and communication. The legal framework is based on strict ethical, technical and legal parameters, can potentially benefit millions of patients from the public health system and state the guidelines for security, privacy, confidentiality and integrity of patient data. The registered medical professional has the autonomy to decide whether to use or refuse telemedicine, indicating face-to-face care if considered necessary. The face-to-face medical consultation remains the gold standard and the autonomy is limited to the principles of beneficence and non-maleficence of the patient and in line with ethical and legal precepts. In addition, formal acceptance must be provided from the patient side.

The data management becomes a critical aspect of this. Several aspects such as data custody, handling, integrity, veracity, confidentiality, privacy, irrefutability and guarantee of professional secrecy of information are necessary to be address. In addition, the physician identity must be verified with digital signature, issued by internationally accredited institutions and the data protection must be in compliance with the requirements from the data protection law. This brings additional critical responsibility for the Electronic Health Record Systems (EHRS) to keep reliable and records with the proper security level to meet the standards of representation, terminology and interoperability.

3 Challenges for Technology Adoption and Maturity

The high inter-patient variability of symptoms and severity of the Coronavirus disease influences multiple aspects, such as resources allocation, patient selection for clinical trials, and individualized strategies for treatment, including vaccination. Variability aspects include a variety of demographics and clinical variables, including geographic and social/economic characterization, biological aspects (age, sex, race), previous diagnostic of comorbidities, and several studies are also identifying genetic aspects, and immune system status and capacity to respond to the disease [?].

This scenario creates significant impact not only on the design and application of clinical protocols and processes, but also on the development of computerized systems for diagnostic support and treatment, which should obtain satisfactory and acceptable performance during the modeling and testing phases, but mainly needs to be validated and obtain the necessary maturity for effective implementation on clinical premises for supporting the decision making process related to the patients.

There is a significant gap between academia and market. The results obtained in the academic environment and published in scientific conferences and journals sometimes are lost on the way to be launched as products in the market, with proper testing, maturity, and problem-solving modeling. This gap should not be considered as a negative aspect, in general. It is, actually, part or the process. The development of specialized clinical support applications are a result of many technical advances which first versions were published as academic works, until moving to the maturity level to become a product.

Nevertheless, during the COVID-19 pandemic, the focus of helping the patients in minimizing their risks or saving their lives, created the necessity to integrate with the clinical practices to support medical decision solutions still in preliminary stages, most of the times based on Artificial Intelligence (AI) and Machine Learning (ML) advanced models and techniques. Some of them could perform a satisfactory role, but on the other hand a significant number could not. One interesting path to provide a classification of these systems could be adopting the classification scale Technology Readiness Levels (TRL's) to these systems, following specific strategies for efficient definitions, given the specificity of this technical area [11]. An introduction with comments regarding common challenges are presented in the following subsection.

3.1 AI Systems and Technology Readiness Challenges

The development of computerized diagnostic support systems based on artificial intelligence techniques is significantly increasing, from classification systems based on neural networks [12] to more advanced and complex models based on Deep Learning [13] networks. According to the most recent developments on the software industry, the Joint Research Center (JRC) of the European Commission published a comprehensive report proposing a methodology to categorize and assess several AI research and development technologies, by mapping them onto Technology Readiness Levels (TRL) (representing their maturity and availability) [14].

Besides the readiness level, the aspect of Generality is a key element to be recognized when considering the analysis of AI systems, since generality is a measure of capacity or performance evaluation of these models. So the level assignment will be directly related to the level of specialization of a proposed solution. For example, the TRL for an AI tool proposed for the analysis of medical images should be clearly identified not only according to which kind of image (which sometimes will differ according to the type of equipment, supplier and even hardware version), but also which specific disease classification or application it is designed to address. If the AI system is a detector, it will detect anomalies in the collected image or set of images, such as a tumor. If the proposed intelligent solution is an advanced classifier, it may have different levels of readiness for a segmentation task, which will allow allow classification of virtually an infinite number of subgroups.

The list of nine TR Levels, with a proper definition originally proposed by National Aeronautics and Space Administration of the United States of America (NASA) [15] and a contextualized application in Machine Learning systems and eventual challenges that the system may face is presented in Table 1.

The next step for AI systems is to align expectations with their end users, the medical and clinical staff, to prove its point and become a necessary tool, with a clear definition of not acting as a substitute of professional specialists, but as an advanced tool to support these specialists to provide better decisions and improve the patients diagnosis or outcome.

4 Current Trends

As a relevant trend, the integration of two areas was selected to represent the continuous impacts of the long term results of the pandemic: the challenge of a significant number of patients living with persistent symptoms as a consequence of the disease, which is classified as Long COVID; and the definition of personalized approaches and multiple biometric data together with the use of Computerized Systems based n Artificial Intelligence to establish a growing area of Precision Medicine, with a focus on COVID-19 and Long Covid patients.

Table 1 xxxx

Level	Definition	Contextualization in ML
TRL1	Basic principles observed	Scientific publications and studies of a machine learning technique and its applications. Not considered a solution for real-life scenarios
TRL2	Technology concept formulated	Implementation of preliminary data analytic approaches. This is also a preliminary phase with no possibility of clinical applications
TRL3	Experimental proof of concept	AI Components, such as different classifiers, implemented but not integrated as a system. Prior to testing in lab environment. Limited solution for practical application
TRL4	Technology validated in the lab	Solution with AI algorithms validated in a laboratory limited environment, normally using previously available data. Still not possible to use in a real-life environment and most of the times already demands a significant amount of resources to create the testing environment
TRL5	Technology validated in a relevant environment	Validation of the AI system in an environment closer to the real application and integration of different modules. This is probably the most common phase of AI systems with the intention to become a commercial application
TRL6	Technology demonstrated in a relevant environment	The AI solution is implemented in a simulated environment or advanced lab. The path between validation (TRL5) and demonstration is sometimes interrupted with the lack of possible external application of the system for demonstration
TRL7	System prototype demonstration in operational environment	The intelligent system is implemented as a prototype in one operational environment, such as a clinic or hospital. Moving from relevant (TRL5 and TRL6) to operational environment (TRL7) is a very difficult step because of multiple requirements such as legal aspects, regulatory frameworks and the significant necessary investment
TRL8	System complete and qualified	The AI system performance is tested and formally approved in real life environment. In this level there is still a long way to achieve a smooth user adoption of the AI system in the operational environment
TRL9	Actual system proven in operational environment	System fully adopted and and AI support for decision making becomes part of the clinical processes and protocols. The practical use of the system will determine if it becomes a useful tool or just one additional functionality not used in a daily basis

4.1 Precision Medicine for COVID-19 and Long COVID Patients

There are multiple statistics to estimate the incidence and prevalence of Long Covid, varying from 9%, which can already be considered a high number, to 35% for certain groups of patients, which, if confirmed with more consistent data analysis and epidemiology models, will configure as a public health challenge itself.

Precision medicine (formerly known as personalized medicine) is a growing area of research related to health sciences [16]. Its development has a straight connection with collecting large amounts of patient data, such as new biomedical information beyond signs and symptoms, which may consider social-economic, clinical data and observations, genetic tests, among others [17]. Understanding of the mechanisms underlying the development of disorders is of urgent necessity, and the developments that have been taking place in recent years have provoked paradigm shifts in the understanding of Medical areas, in addition to better clinical benefits and a significantly lower cost [18].

In this field there is still significant space for exploration, computerized supporting systems have a promising role by providing tools that allow the analysis of large amounts of data recruited from numerous factors that combine to trigger disorders, which are known to be determined by multi-factorial conditions [19]. In this sense, artificial intelligence assumes a relevant role, as it appropriates the use of data-driven algorithms, that is, they progressively improve as receive training on top of the information that is being increased [20]. That opens up a range of possibilities for a better understanding of health issues and disorders, especially for severe prognosis of COVID-19 for some patients and the recent therapeutics and protocols to support patients with Long Covid.

In this scenario, dealing with large amount of structure and unstructured data, from multiple types (images, audios, videos, questionnaires, laboratory results, etc.) the development of computerized systems to support medical diagnosis and patient treatment becomes an essential step to address the multiple existing outcomes and types of treatment, especially with more transversal and long term clinical trials and following up clinical interventions.

5 What to Expect from This Book

This book aims to discuss developments on Computerized Systems for the diagnostic and treatment of the COVID-19 pandemic, so the discussion here will be focused on multiple solutions and approaches using computational tools to support medical applications, including computational vision for medical imaging, exploratory data analysis of several clinical data of different groups or patients, biosignals processing, considering the adoption of different artificial intelligence models, according to the area of study.

In the first part, efficient techniques to accurately diagnose and classify COVID-19 and its impacts on the patients' lungs, two types of medical imaging are considered: X-Ray and CT-Scan. The image quality and potential analysis is not under comparison, since the set of images provided in a computerized tomography is enormously more precise than one X-Ray of the chest. The lower operational and financial cost, feasibility of execution for hospitalized patients, and simplified clinical operations, are the reasons to highlight the importance of using the X-Ray, and consequently, develop computerized systems to support medical decision-making on these images. Artificial Intelligence models based on Deep Learning approaches are presented in a didactic

In the second part of the book, the focus is on the analysis of clinical and cardiovascular data of COVID-19 patients. Together with multiple blood analysis and other bio-markers, the Electrocardiogram (ECG) of infected patients, classified in different degrees of severity, were collected. Two different strategies were followed. Firstly, the ECG signal is considered and multiple approaches for a classification system are proposed. Secondly, the Heart Rate Variability (HRV), which is determined by the variation of interval between sequential heartbeats as a reflex of the Autonomous Nervous System (ANS), is used as input for a group of different artificial intelligence classifiers.

As a final discussion, an exploratory data analysis of multiple data from pregnant women diagnosed with Covid and with persistent symptoms is presented. The condition known as Post COVID-19 or long COVID, can affect anyone exposed to the new Corona Virus (SARS-CoV-2), regardless of age or severity of original symptoms. According to the WHO, it is defined as the continuation or development of new symptoms 3 months after the initial infection, with these symptoms lasting for at least 2 months with no other explanation [21]. Since this affects a significant number of patients, it becomes extremely necessary to analyze, understand and cope with the new challenges rising from this issue. Some common symptoms are cognitive impairment, fatigue, or shortness of breath, individuals report more than two hundred different issues, impacting of their daily health conditions.

References

1. Lobo Marques JL, Nauber Bernardo Gois F, Xavier-Neto J, Fong SJ (2021) Predictive models for decision support in the COVID-19 crisis. SpringerBriefs in applied sciences and technology. Springer International Publishing, Cham
2. Lobo Marques JA, Fong SJ (eds) (2022) Epidemic analytics for decision supports in COVID19 crisis. Springer International Publishing, Cham
3. Lobo Marques JA, Nauber Bernardo Gois F, Xavier-Neto J, Fong SJ (2021) Nonlinear prediction for the COVID-19 data based on quadratic Kalman filtering. In: Lobo Marques JA, Nauber Bernardo Gois F, Xavier-Neto J, Fong SJ (eds) Predictive models for decision support in the COVID-19 crisis. SpringerBriefs in applied sciences and technology. Springer International Publishing, Cham, pp 55–68

4. Lobo Marques JA, Nauber Bernardo Gois F, Xavier-Neto J, Fong SJ (2021) Forecasting COVID-19 time series based on an autoregressive model. In: Lobo Marques JA, Nauber Bernardo Gois F, Xavier-Neto J, Fong SJ (eds) Predictive models for decision support in the COVID-19 crisis. Springer briefs in applied sciences and technology. Springer International Publishing, Cham, pp 41–54
5. Mayo Clinic (2020) Covid-19 and related vaccine development and research. https://www.mayoclinic.org/coronavirus-covid-19/history-disease-outbreaks-vaccine-timeline/covid-19. Accessed: 2023-01-05
6. Voshaar T, Randerath W, Bauer T, Geiseler J, Dellweg D, Westhoff M, Windisch W, Schünhofer B, Kluge S, Lepper PM, Pfeifer M, Ewig S (2020) Positionspapier zur praktischen umsetzung der apparativen differenzialtherapie der akuten respiratorischen insuffizienz bei covid-19 [position paper for the state of the art application of respiratory support in patients with covid-19 - german respiratory society]. Pneumologie 74(6):337–357
7. Esquinas AM, Mazza M, Fiorentino G (2022) Elmo helmet for cpap to treat covid-19-related acute hypoxemic respiratory failure outside the icu: aspects of/comments on its assembly and methodologyauthors' replypatient self-inflicted lung injury and positive end-expiratory pressure for safe spontaneous breathingelmo 1 0: a helmet interface for cpap and high-flow oxygen deliveryprotecting healthcare workers from sars-cov-2 infection practical indications. J Bras Pneumol 48(2)
8. Lino JA, Menezes DGA, Soares JB, Furtado V, Soares Júnior L, Farias MDSQ, Lima DLN, Pereira EDB, Holanda MA, Tomaz BS, Gomes GC (2022) Elmo, a new helmet interface for cpap to treat covid-19-related acute hypoxemic respiratory failure outside the icu: a feasibility study. J Bras Pneumol 48(1)
9. School of Public Health of Ceara Brazil (2022) Elmo respiratory support project - covid-19. https://clinicaltrials.gov/ct2/show/NCT04470258. Accessed: 2023-01-06
10. Federal Council of Medicine Brazil (2022) Resolution 2.314 - telemedicine regulatory act. https://www.in.gov.br/web/dou/-/resolucao-cfm-n-2.314-de-20-de-abril-de-2022-397602852. Accessed: 2023-01-10
11. Gilligan-Lee CM, Visnjic A et al Lavin A (2022) Technology readiness levels for machine learning systems. Nat Commun 13(1):6039
12. Lobo Marques JA, Nauber Bernardo Gois F, Paulo do Vale Madeiro J, Li T, Fong SJ (2022) Chapter 4 - Artificial neural network-based approaches for computer-aided disease diagnosis and treatment. In: Kumar Bhoi A, de Albuquerque VHC, Naga Srinivasu P, Marques G (eds) Cognitive and soft computing techniques for the analysis of healthcare data. Intelligent data-centric systems. Academic Press, pp 79–99
13. Lobo Marques JA, Nauber Bernardo Gois F, Aryel Nunes da Silveira J, Li T, Fong SJ (2022) Chapter 5 - AI and deep learning for processing the huge amount of patient-centric data that assist in clinical decisions. In: Kumar Bhoi A, de Albuquerque VHC, Naga Srinivasu P, Marques G (eds) Cognitive and soft computing techniques for the analysis of healthcare data. Intelligent data-centric systems. Academic Press, pp 101–121
14. Gomez Gutierrez E, Martinez Plumed F, Hernández-Orallo J (2020) Ai watch: assessing technology readiness levels for artificial intelligence. Publications office of the European union, p 72
15. NASA (2012) Technology readiness level. https://www.nasa.gov/directorates/heo/scan/engineering/technology/technology_readiness_level. Accessed: 2023-01-05
16. Uziel D, de Negri F (2020) O que é medicina de precisão e como ela pode impactar o setor de saúde? http://repositorio.ipea.gov.br/handle/11058/9970. Accessed: 2023-01-14
17. Gesine H, von Mutius E, Kopp MV, Künig IR, Fuchs O (2017) What is precision medicine? Eur Respir J 50(4):1700391
18. Zhao Z, Fernandes BS, Quevedo J (2022) Fostering precision psychiatry through bioinformatics. Braz J Psychiatry 44(2):119–20

19. Li JL, Mondello S, Nokkari A, Razafsha M et al, Alawieh A, Zaraket FA (2012) Systems biology, bioinformatics, and biomarkers in neuropsychiatry. Front Neurosci 6
20. Malik YK, Singh S, Gupta R, Ray A, Bhardwaj A (2022) Artificial intelligence and psychiatry: an overview. Asian J Psychiatr 70:103021
21. World Health Organization (WHO) (2022). Post covid-19 condition - long covid. https://www.who.int/europe/news-room/fact-sheets/item/post-covid-19-condition. Accessed: 2023-01-14

Lung Segmentation of Chest X-Rays Using Unet Convolutional Networks

Bruno Riccelli dos Santos Silva, Paulo Cesar Cortez, Rafael Gomes Aguiar, Tulio Rodrigues Ribeiro, Alexandre Pereira Teixeira, Francisco Nauber Bernardo Gois, and Joao Alexandre Lobo Marques

Abstract The gold standard to detect SARS-CoV-2 infection consider testing methods based on Polymerase Chain Reaction (PCR). Still, the time necessary to confirm patient infection can be lengthy, and the process is expensive. On the other hand, X-Ray and CT scans play a vital role in the auxiliary diagnosis process. Hence, a trusted automated technique for identifying and quantifying the infected lung regions would be advantageous. Chest X-rays are two-dimensional images of the patient's chest and provide lung morphological information and other characteristics, like ground-glass opacities (GGO), horizontal linear opacities, or consolidations, which are characteristics of pneumonia caused by COVID-19. But before the computerized diagnostic support system can classify a medical image, a segmentation task should usually be performed to identify relevant areas to be analyzed and reduce the risk of noise and misinterpretation caused by other structures eventually present in the images. This chapter presents an AI-based system for lung segmentation in X-ray

B. R. dos Santos Silva · P. Cesar Cortez · R. Gomes Aguiar · T. Rodrigues Ribeiro · A. Pereira Teixeira
Laboratório de Engenharia e Sistemas de Computação, Universidade Federal do Ceará, Campus do Pici, Fortaleza, Brazil
e-mail: bruno@lesc.ufc.br

P. Cesar Cortez
e-mail: cortez@lesc.ufc.br

R. Gomes Aguiar
e-mail: rafael.aguiar@ufu.br

T. Rodrigues Ribeiro
e-mail: tuliorr@edu.unifor.br

A. Pereira Teixeira
e-mail: alexandreapt@gmail.com

F. N. Bernardo Gois · J. A. Lobo Marques (✉)
Laboratory of Applied Neurosciences, University of Saint Joseph, Estrada Marginal da Ilha Verde, 14-17, Macao SAR, China
e-mail: alexandre.lobo@usj.edu.mo

F. N. Bernardo Gois
e-mail: nauber.gois@usj.edu.mo

J. A. Lobo Marques and S. J. Fong (eds.), *Computerized Systems for Diagnosis and Treatment of COVID-19*, https://doi.org/10.1007/978-3-031-30788-1_2

images using a U-net CNN model. The system's performance was evaluated using metrics such as cross-entropy, dice coefficient, and Mean IoU on unseen data. Our study divided the data into training and evaluation sets using an 80/20 train-test split method. The training set was used to train the model, and the evaluation test set was used to evaluate the performance of the trained model. The results of the evaluation showed that the model achieved a Dice Similarity Coefficient (DSC) of 95%, Cross entropy of 97%, and Mean IoU of 86%.

1 Introduction

Some ways of detecting the SARS-CoV-2 infection include Nucleic acid amplification tests and antigen ones. On the one hand, techniques based on Polymerase Chain Reaction (PCR) are the gold standard testing method, but the time necessary to confirm patient infection can be lengthy [1, 2]. On the other hand, X-Ray and CT scans play a vital role in the auxiliary diagnosis process. Hence, a trusted automated technique for identifying and quantifying the infected lung regions would be pretty advantageous [3].

Comparing CT scans with X-Ray, we note that the second one is cheaper and produces minimal radiation to the patient. Moreover, X-ray machines may provide two-dimensional images of the patient's chest in most radiological laboratories, and hospitals [3]. This acquisition method provides lung information about the morphological area and allows a more detailed investigation of various clinical conditions like emphysema and pneumothorax. It is noticeable that some lung characteristics like ground-glass opacities (GGO), horizontal linear opacities, or consolidation change during infection by SARS-COV-2 [3–5].

Artificial intelligence (AI) is a growing area that allows a machine "learn" behaviors and infer unseen data. Machine Learning is a subset of AI associated with human intelligence, improving data analyses through computational algorithms. After sufficient repetitions and modification of the algorithm, the algorithm can take an input and predict an output [6].

Convolutional Neural Networks (CNNs) are commonly used in classification tasks, where the output of an image is a single class label. However, in biomedical image processing tasks, the intended result must include localization, i.e., a class label must be assigned to each pixel of the processed image. These types of Artificial Neural Networks (ANN) became very popular for image processing, with a large number of applications in the area of medical imaging.

In this context, Ronneberger, Fischer, and Brox (2015) proposed a new architecture called U-net, which was an improvement of the Fully Convolutional Networks for Semantic Segmentation or Fully Convolutional Networks (FCN) developed by Long, Shelhamer, and Darrell (2014). In addition, the authors modified this architecture to work with few training images and produce more accurate segmentations.

The work proposed by [7] presented a new approach for brain tumor segmentation using Deep Neural Networks using Magnetic Resonance Images (MRI). The authors applied a CNN to 33×33 images of BraTS 2013 dataset. The collected data consists of 30 brain images with high-grade and low-grade gliomas. They presented a two-pathway CNN architecture called TwoPathCNN to generate feature maps and segment the tumor regions. Dice coefficient, sensitivity, and specificity were applied to evaluate the segmented tumors. They obtained up to 88% of the Dice coefficient in the tumor region.

In another work, proposed by [8], nine classifiers (Naive Bayes, K-Nearest-Neighbors, Generalized Linear Models, Gradient Boosting, AdaBoost, Random Forests, Extra Tree Forests, Tuned Extra Tree Forests, and CNN) were evaluated through 37 MRI datasets of stroke patients for lesion segmentation. The intensity feature, the weighted local mean, the 2D center distance, and the local histogram were used for feature extraction. They evaluated the classifiers through Dice Coefficient, Average symmetric surface distance, and Hausdorff distance metrics. Random Forests and CNN approaches achieved the best segmentation results.

Another recent image segmentation approach is Explainable Artificial Intelligence (XAI). In [9], the authors used XAI to extract a pixel-wise binary segmentation from the Layer-wise Relevance Propagation (LRP) output, explaining the decision of a classification network. They used the VGG-11 architecture with 128 neurons in the fully connected layers. Two datasets were used: Cracks in sewer pipes for classification of damaged and undamaged pipe surfaces and cracks in magnetic tiles with either a crack or a blow-hole. They compared the proposed segmentation technique with U-net through Intersection over Union, precision, and recall metrics, obtaining competitive results.

Deep learning models, such as U-nets, have been used to identify early tumors that can lead to cancer in different organs, such as the lungs. Unfortunately, lung cancer is the first among men and the second among women, according to OMS. In most cases, the disease is diagnosed in an advanced stage, reducing the survival rate.

A U-net-based network was proposed by [10] using the PASCAL VOC 2012 dataset for medical image segmentation. They used (480,480) images for 320 epochs during the training phase. For evaluation of the proposed network, the authors used Promise12, Chaos, and NERVE datasets that consist of an MRI, Computed Tomography, and ultrasound, respectively. In addition, they used the dice coefficient and the mean intersection over union metrics for evaluation.

Another U-net architecture proposed was presented in [11]. The authors contributed with a U-net enhancement called Unet++. They compared with U-net standard across medical segmentation problems such as nodule segmentation and CT scans of the chest and abdomen. The U-net++ presented significant results over U-Net in terms of Intersection over Union metric.

Considering the possibility of successfully adopting CNN models for image segmentation, this chapter will describe and discuss an intelligent lung segmentation system from X-Ray images based on a proposed U-net CNN Model. The metrics considered for performance evaluation are the Dice Similarity Coefficient (DSC), Mean Intersection over Union (IoU) and Cross Entropy. The system will perform the segmentation step of a complete solution for X-Ray diagnosis and analysis for COVID-19 patients.

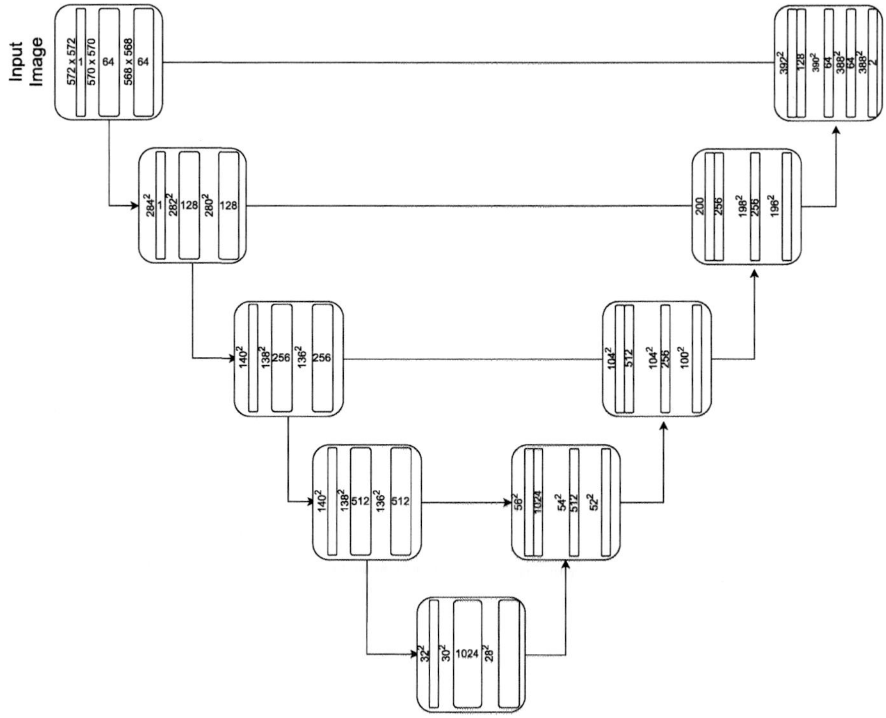

Fig. 1 U-net architecture

2 Materials and Methods

This Section presents our methodology details, describing the procedure for achieving the desired results (Fig. 1).

2.1 U-Net Convolutional Networks

Developed by Olaf Ronneberger, Philipp Fischer, and Thomas Brox, the U-net is a convolutional neural network used for several segmentation purposes, such as medical, cell nuclei in biology areas, industrial inspection applications, satellite images, and self-driving car tasks. The architecture consists of convolutional layers and pooling layers used for detection patterns and resizing the feature maps generated in the convolutional layers. It has two networks: an encoder and a decoder. The first aim is to compress the input image to preserve the most important information for the segmentation task. After that, these generated feature maps are passed to the decoder

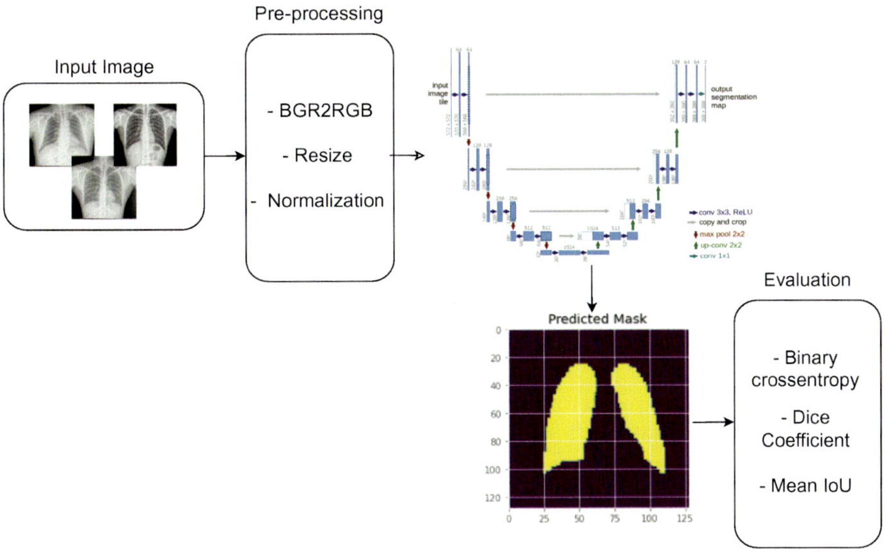

Fig. 2 Methodology flowchart for this chapter

network, which uses it to generate the final output. Figure 2 presents the U-Net architecture.

2.2 *Proposed System*

Figure 2 shows a high-level overview of the methodology employed. First, we have the X-ray images from our collected data in the input phase. Then, the pre-processing step consists of standard image treatment techniques like RGB conversion, resizing, and Normalization. Note that these procedures are necessary for the transfer learning input format. After that, we extract features from the pre-trained U-net architecture.

The U-Net architecture consists of an encoder path (contraction) and a decoder path (expansion), which can be defined as:

- Encoder (left part of the "U")—encodes the image into an abstract representation of its features by applying a sequence of convolutional blocks that gradually reduce the height and width of the representation, but an increasing number of channels that correspond to the image features.
- Decoder (right part of the "U")—decodes the image representation into a binary mask by applying a sequence of ascending convolutions that gradually increases the height and width of the representation to the size of the original image and decreases the number of channels to the number of classes we are segmenting.

The network uses only the valid part of each convolution. It has no fully connected layers, i.e., the segmentation map contains only those pixels for which the entire context is available in the input image. This strategy allows a block overlay strategy to segment arbitrarily large images. Furthermore, the U-net network performs semantic image segmentation, i.e., labels each pixel of an image with a corresponding class of what is being represented. Thus, we expect to output a pixel-level image classification.

2.3 Evaluation Metrics

We evaluated the U-net network model segmentation performance using the Dice similarity coefficient (DSC), Cross entropy, and Mean Intersection over Union (IoU). DSC measures the proportion of the intersection between the voxels extracted by the segmentation model and the voxels in the ground truth [12]. The DSC can be defined as:

$$DSC = \frac{2|o(P) \cap Y|}{|o(P)| + |Y|} \tag{1}$$

where o(.) is the OTSU binarization method, Y represents the gold standard of lung mask, and P represents the predicted probability map.

The cross-entropy is the average number of bits needed to encode data from a source with distribution p. When we use model q to predict the probability map, it is used as a similarity metric to tell how close one distribution of random events is to another and is used for classification and segmentation. The cross-entropy is defined as:

$$CrossEntropy = -\sum_{i}^{C} t_i * log(p_i) \tag{2}$$

where t_i is the gold standard of the lung mask and p_i is the probability of the ith mask.

The Mean Intersection over Union can be defined as:

$$MeanIoU = (1/n) * sum(IoU(predicted_i, groundtruth_i)) \tag{3}$$

where n is the number of images in the train/test set, IoU($predicted_i, groundtruth_i$) is the Intersection over Union (IoU) of the i-th predicted bounding box and the i-th ground truth bounding box for a single image.

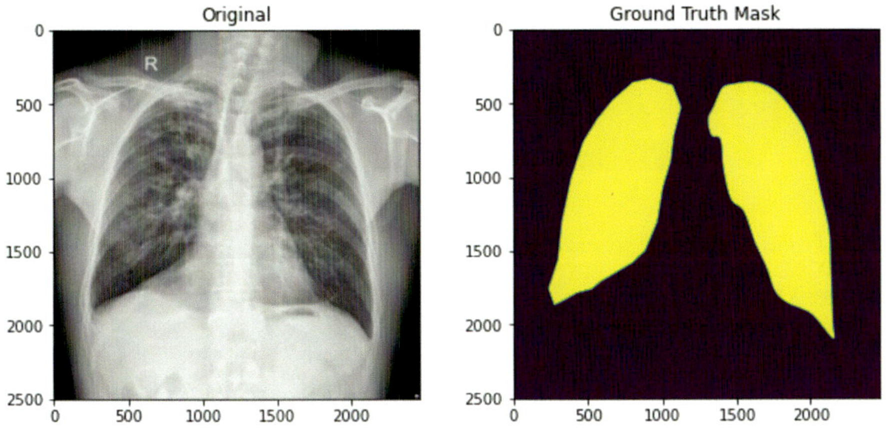

Fig. 3 Original and ground truth mask of the adopted dataset

2.4 Dataset

X-ray images in this data set have been acquired from the tuberculosis control program of the Department of Health and Human Services of Montgomery County, MD, USA, and Shenzhen People's Hospital, Guangdong Medical College, Shenzhen, China [13, 14]. The Chest X-rays are collected from out-patient clinics and were captured as part of the daily routine using Philips DR Digital Diagnose systems. This set contains 800 posterior-anterior x-rays, of which 406 x-rays are normal and 394 x-rays are abnormal with manifestations of tuberculosis. All images are identified and available in DICOM format. Figure 3 illustrates an original exam and its ground truth mask.

Note that the segmentation task is essential for detecting, for example, nodules or predicting the lung sequelae percentage.

2.5 Experimental Design

In this Section, we describe our experimental design applied in this Chapter. We first resize the input image to (256, 256, 3) because the U-net has this image segmentation requirement. After that, we convert the images to RGB and normalize the input data. Therefore, the output is the predicted mask of the chest.

We performed an 80/20 train test split to divide data for the train and evaluation test. The U-net architecture was applied with [32,64,128,256] filter sizes. The Adam optimizer was adopted during 100 epochs. Moreover, we used an early stop of ten epochs tolerance to prevent overfitting. The evaluated metrics were DSC, binary cross-entropy, and mean IoU. We applied a batch size of 16 in this experiment.

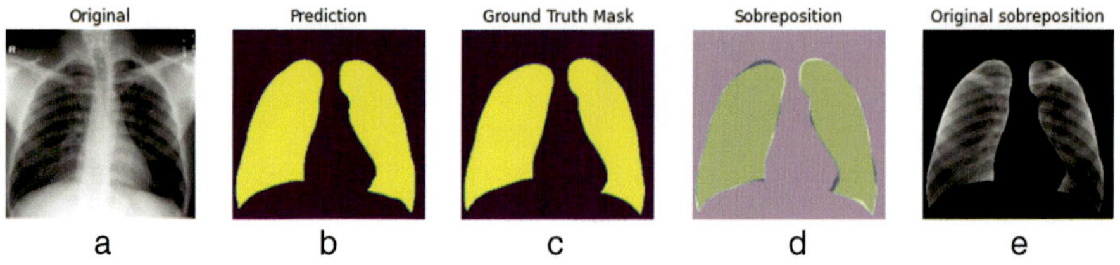

Fig. 4 Prediction mask and ground truth comparison

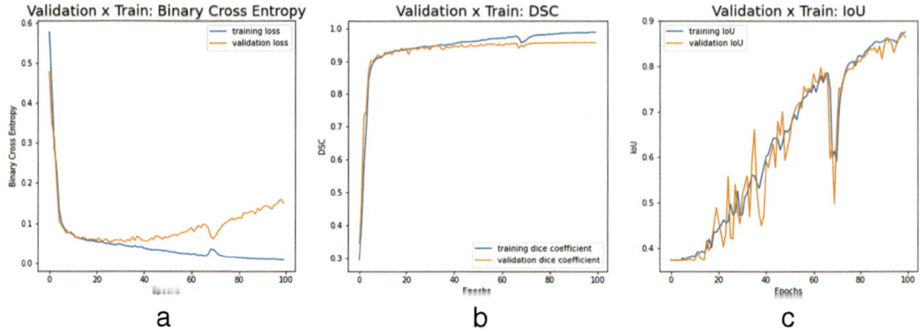

Fig. 5 Cross entropy and DSC evaluation metrics through the epochs

3 Results and Discussions

In this section, we present and discuss the main results achieved in this chapter. Figure 4 illustrates an example of successful segmentation from this experimental design employed in this chapter.

This example image has a similarity coefficient greater than 97%. This result indicates that U-net architecture performs well for chest segmentation purposes. Moreover, the mean DSC obtained was 97%, considering all evaluated images.

Figure 5 presents the train and validation metrics along the evaluated epochs.

The cross-entropy is reducing along the epochs, and the coefficient dice is growing. This behavior is expected because they are non-proportional metrics and means that our model can segment the chest from X-ray images. Moreover, the IoU is stabilizing over the epochs for training and validation sets, achieving results above 85%.

Regarding the test evaluation, 141 new chest x-ray images were submitted to our segmentation methodology, achieving 95%, 97%, and 86% for DSC, Cross entropy, and mean IoU, respectively.

Moreover, Fig. 6 illustrates the DSC metric histogram for this evaluation. We can see that the average of the Dice coefficient was higher than 95.00%, obtaining a good performance on the test data.

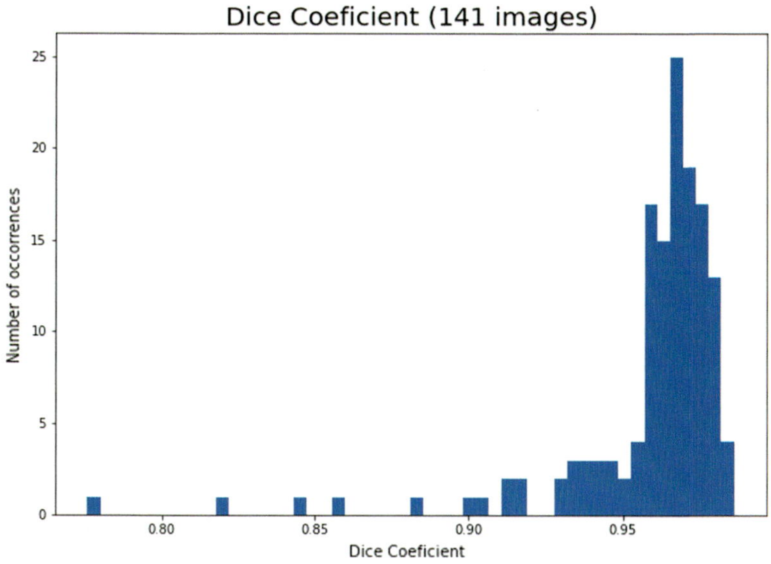

Fig. 6 Dice coefficient

3.1 Clinical Examples

In this section, we discuss some clinical examples evaluated in test split. The examples discussed here are actual exams not submitted to U-net architecture in the training phase, presenting a real analysis of the performance of the proposed solution.

In Fig. 7, is presented the general modeling for the experiment under analysis. Different clinical cases are presented following this approach to allow a better comprehension from the reader.

Figure 8 illustrates the prediction, the ground truth mask, and the superimposition between prediction and ground truth related to the original image collected in the dataset. The DSC coefficient and the Mean IoU for this case are 97% and 96%, respectively.

Note that the prediction subfigure has a segmentation error on the right lung. This behavior may occur because the heart region resembles similar-intensity lung thresholds. Moreover, the heart beating can produce motion artifacts and lead to errors in the segmentation. Another approach for dealing with this issue is heart segmentation or combining other techniques, such as MRI and ultrasound, to obtain more differences between the two organs.

Figure 9 illustrates the second clinical example result of the methodology applied in this Chapter. The DSC coefficient and the Mean IoU for this case are 95% and 94%, respectively.

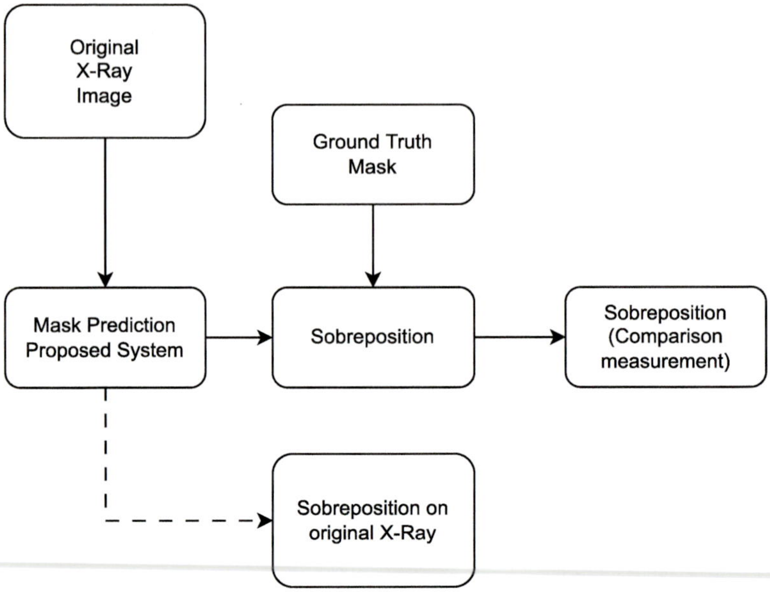

Fig. 7 Clinical example 1

Fig. 8 Clinical example 1

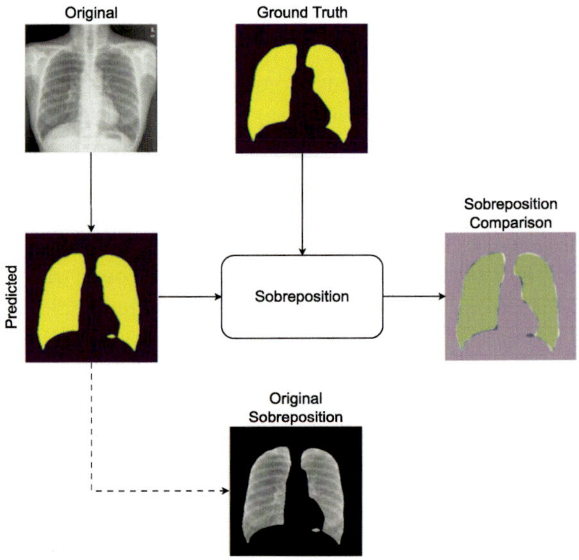

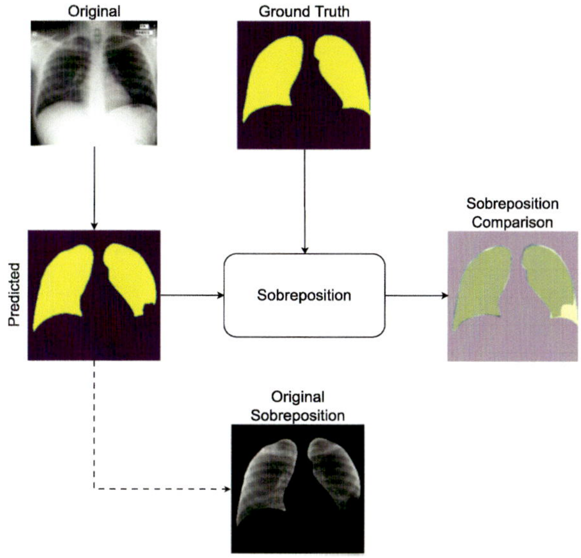

Fig. 9 Clinical example 2

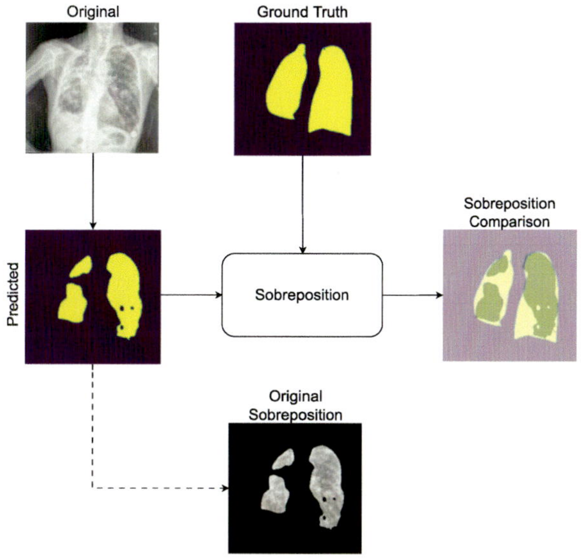

Fig. 10 Clinical example 3

The right lung was not segmented entirely. This behavior is explained because the lower right lung is at the edge of the image, and most of the samples in the dataset have a farther frame between the chest and the x-ray equipment.

Figure 10 presents the third clinical case, and we highlight a high error in the segmentation result. The left lung was not recognized in the segmentation process. The DSC coefficient and the Mean IoU for this case are 77% and 95%, respectively.

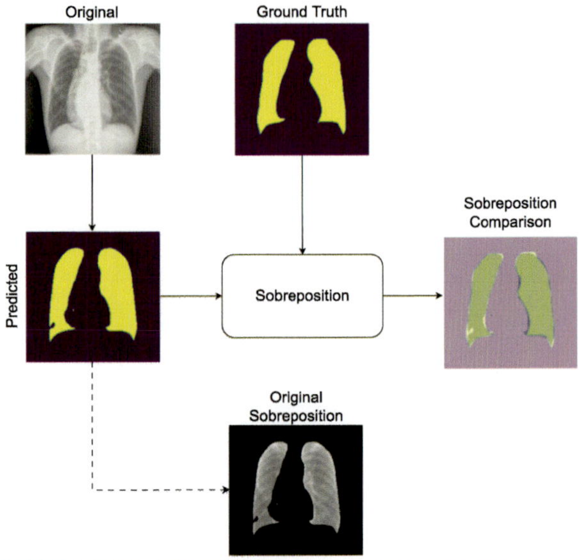

Fig. 11 Clinical example 4

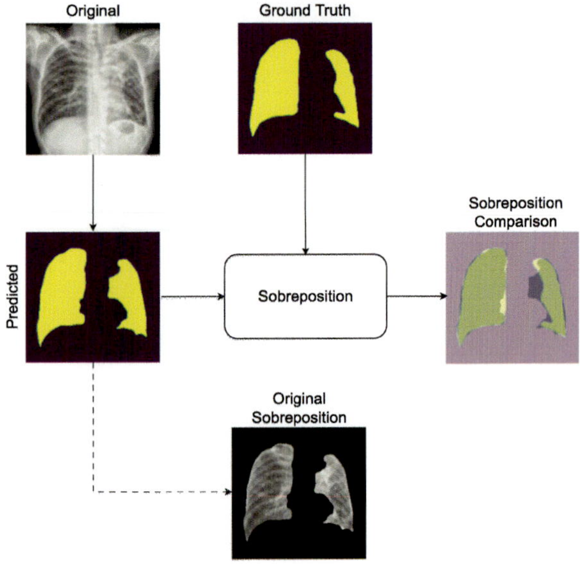

Fig. 12 Clinical example 5

Here we note the lateral posture at the moment of the exam. Moreover, the image quality is poor, difficulting to separate the lung thresholds.

A good segmentation is presented in Fig. 11. The DSC coefficient and the Mean IoU for this case are 95% and 94%, respectively. In the left lung, there are some pixels not seen by the U-net. Note that some artifacts, like the heart, may produce errors due to the similar threshold compared to the lung region.

Figure 12 illustrates the last clinical example. The left lung was correctly segmented. However, the right one is significantly compromised by lesions and secretions. Thus, the predicted and the ground truth masks have a small area compared to a normal lung. The DSC coefficient and the Mean IoU for this case are 91% and 89%, respectively.

It is worth noting that there are a few reasons why lung segmentation may produce errors. Some possible reasons are, for example, the lack of standardization in the acquisition of x-ray images, related to the positioning of the body, and poor image quality. Moreover, the heart artifact can produce errors in the segmentation due to the similar-intensity lung thresholds.

4 Conclusions

The accurate computer-aided analysis of X-ray images for identifying COVID-19 or other lung diseases, is a challenging task. However, it has the potential to greatly aid in the medical decision-making process for diagnosis and patient management.

In this chapter, presented an artificial intelligence-based lung segmentation system for X-ray images using a U-net CNN model. The cross-entropy, dice coefficient, and Mean IoU were applied to evaluate the employed architecture in unseen data. The results showed that the U-net presented promising results for lung segmentation purposes, achieving 95%, 97%, and 86% for DSC, Cross entropy, and mean IoU, respectively. However, it may produce errors due to a lack of standardization in x-ray image acquisition, poor image quality, and the heart artifact, which can produce similar-intensity lung thresholds.

Furthermore, this Chapter did not exhaust the possibilities of researching COVID-19 segmentation. For future works, we can segment Covid-19 using another CNN-based architectures, segment lesions and predict the lung sequelae percentage.

References

1. Gopatoti A, Vijayalakshmi P (2022) Optimized chest X-ray image semantic segmentation networks for COVID-19 early detection. J X-Ray Sci Technol, pp 1–22
2. Zhang F (2021) Application of machine learning in CT images and X-rays of COVID-19 pneumonia. Medicine 100
3. Saood A, Hatem I (2021) COVID-19 lung CT image segmentation using deep learning methods: U-Net versus SegNet. BMC Med Imaging 21:1–10
4. Sathi S, Tiwari R, Verma S, Kumar Garg A, Singh Saini V, Kumar Singh M, Mittal A, Vohra D (2021) Role of chest x-ray in coronavirus disease and correlation of radiological features with clinical outcomes in Indian patients. Can J Infect Dis Med Microbiol 2021
5. Vaz Rodrigues L, Martins Y, Guimaraes C, de Santis M, Marques A, Barata F (2011) Anatomy for the bronchologist: a prospective study of the normal endobronchial anatomic variants. Revista Portuguesa de Pneumologia (English Edition) 17, 5:211–215

6. Helm J, Swiergosz A, Haeberle H, Karnuta J, Schaffer J, Krebs V, Spitzer A, Ramkumar P (2020) Machine learning and artificial intelligence: definitions, applications, and future directions. Curr Rev Musculoskelet Med 13:69–76
7. Havaei M, Davy A, Warde-Farley D, Biard A, Courville A, Bengio Y, Pal C, Jodoin P, Larochelle H (2017) Brain tumor segmentation with deep neural networks. Med Image Anal 35:18–31
8. Maier O, Schrüder C, Forkert N, Martinetz T, Handels H (2015) Classifiers for ischemic stroke lesion segmentation: a comparison study. PloS One 10:e0145118
9. Seibold C, Künzel J, Hilsmann A, Eisert P (2022) From explanations to segmentation: using explainable AI for image segmentation. arXiv:2202.00315
10. Weng Y, Zhou T, Li Y, Qiu X (2019) NAS-Unet: neural architecture search for medical image segmentation. IEEE Access 7:44247–44257
11. Zhou Z, Rahman Siddiquee M, Tajbakhsh N, Liang J (2018) Unet++: a nested u-net architecture for medical image segmentation. In: Deep learning in medical image analysis and multimodal learning for clinical decision support, pp 3–11
12. Zhao C, Xu Y, He Z, Tang J, Zhang Y, Han J, Shi Y, Zhou W (2021) Lung segmentation and automatic detection of COVID-19 using radiomic features from chest CT images. Pattern Recognit 119:108071
13. Jaeger S, Karargyris A, Candemir S, Folio L, Siegelman J, Callaghan F, Xue Z, Palaniappan K, Singh R, Antani S et al (2013) Automatic tuberculosis screening using chest radiographs. IEEE Trans Med Imaging 33:233–245
14. Candemir S, Jaeger S, Palaniappan K, Musco J, Singh R, Xue Z, Karargyris A, Antani S, Thoma G, McDonald C (2013) Lung segmentation in chest radiographs using anatomical atlases with nonrigid registration. IEEE Trans Med Imaging 33:577–590

Segmentation of CT-Scan Images Using UNet Network for Patients Diagnosed with COVID-19

Francisco Nauber Bernardo Gois and Joao Alexandre Lobo Marques

Abstract The use of computational tools for medical image processing are promising tools to effectively detect COVID-19 as an alternative to expensive and time-consuming RT-PCR tests. For this specific task, CXR (Chest X-Ray) and CCT (Chest CT Scans) are the most common examinations to support diagnosis through radiology analysis. With these images, it is possible to support diagnosis and determine the disease's severity stage. Computerized COVID-19 quantification and evaluation require an efficient segmentation process. Essential tasks for automatic segmentation tools are precisely identifying the lungs, lobes, bronchopulmonary segments, and infected regions or lesions. Segmented areas can provide handcrafted or self-learned diagnostic criteria for various applications. This Chapter presents different techniques applied for Chest CT Scans segmentation, considering the state of the art of UNet networks to segment COVID-19 CT scans and a segmentation experiment for network evaluation. Along 200 epochs, a dice coefficient of 0.83 was obtained.

1 Introduction

The use of computational tools for medical image processing are promising tools to effectively detect COVID-19 as an alternative to expensive and time-consuming RT-PCR (Reverse Transcription—Polymerase Chain Reaction) tests [1, 2]. For this specific task, CXR (Chest X-Ray) and CCT (Chest CT Scans) are the most common examinations to support diagnosis through radiology analysis [3]. With these images, it is possible to support diagnosis and determine the disease's severity stage. Computerized COVID-19 quantification and evaluation require an efficient segmentation process. Essential tasks for automatic segmentation tools are precisely identifying the

F. N. Bernardo Gois · J. A. Lobo Marques (✉)
Laboratory of Applied Neurosciences, University of Saint Joseph,
Estrada Marginal da Ilha Verde, 14-17, Macao SAR, China
e-mail: alexandre.lobo@usj.edu.mo

F. N. Bernardo Gois
e-mail: nauber.gois@usj.edu.mo

J. A. Lobo Marques and S. J. Fong (eds.), *Computerized Systems for Diagnosis and Treatment of COVID-19*, https://doi.org/10.1007/978-3-031-30788-1_3

lungs, lobes, bronchopulmonary segments, and infected regions or lesions. However, CT scans also reveal imaging characteristics similar to those of other pneumonia, making it difficult to distinguish between them. In addition, the manual portrayal of lung regions or infected parts is a laborious and time-consuming process frequently impacted by personal bias and clinical expertise.

In recent years, Artificial Intelligence systems based on deep learning networks have been considered for multiple applications. Especially for medical image processing, several tasks are executed with sufficient accuracy. For example, for image segmentation [4]. Other works considered fully-connected CNNs to generate input-resolution segmentation outputs using fractionally strung convolution, commonly known as upsampling or deconvolution [5], achieving Mean Intersection of Union (M-IOU) values of 62.7%, 34.0%, and 39.0%, for the datasets PASCAL VOC, NYUDv2, and SIFT, respectively. In addition, the work claimed that upsampling, a component of the proposed system, was fast and efficient for predictions. Following from these developments, the UNET structure was proposed as an evolution of fully-connected CNNs for medical image processing, focused on the segmentation of relevant structures [6, 7].

As explained, the segmentation task for medical images to classify COVID-19 has similar challenges and is an essential step for the following steps [8]. And deep learning has been widely considered for that task [9, 10]. U-Net, V-Net, and U-Net++ [11, 12] were used for COVID-19 segmentation, along with other frameworks commonly employed for medical image segmentation. In another work, Wang et al. [13] proposed a framework to address the issue of poor image labeling or lack of labeling, considering morphological aspects of Chest images. Also interested in mitigating the same issue, another work [14] used a supervised approach [15]. In addition, another work used a segmentation strategy based on two stages, firstly using the U-net for a first-level segmentation and multiple techniques to segment the relevant areas of the image in detail [16].

As can be seen from the literature presented, the deep learning model called U-Net is a popular approach for the segmentation of medical images due to its effectiveness and sturdiness as a segmentation model. A more detailed understanding of how it works is presented in the following Sections. Many earlier works focused on improving the performance of segmentation at the pixel level [17]. In the SD-UNet proposed in this Chapter, a squeeze-and-attention (SA) module was developed to overcome these obstacles by employing a not fully-squeezed attention channel mechanism to generate non-local spatial attention to the image and fully use the global context information to selectively re-weight the channel features. It can also be viewed as a spatial awareness of pixel grouping. By producing attention convolution, it is possible to create groups of pixels based on multiple spatial characteristics [15].

This chapter presents the state of the art of using UNet networks for the segmentation of Chest CT scans of patients diagnosed with COVID-19 and a segmentation experiment for the model evaluation. Along 200 epochs, a Dice Coefficient of 0.83 was obtained.

2 Background

2.1 Lung CT Scans Characteristics and COVID-19 Diagnosis

Computerized Tomography (CT) is a medical imaging technology used to detect detailed characteristics in the interior organs' structure [18]. In addition to an X-Ray, this procedure employs three-dimensional computer vision technology for a comprehensive investigation of organs. Both CT scans and X-rays capture images of internal organs and structures. Nevertheless, X-rays are suitable to overlap structures due to the unique image generation technology. In CT images, the radiology emissions with multiple incidences result in high-quality images.

The scientific literature indicates that pneumonia is the root cause of lung damage in COVID-19 patients, normally bilateral, [19], which makes these legions similar to pneumonia-related diseases [20]. Some common patterns of COVID-19 lung images are found. First, the most common are Ground-Glass Opacities (GGOs) and are explained in the remaining of this Section. Other detectable patterns are "crazy-paving" regions, consolidations, thickening of interlobular septa, reticular pattern, mixed pattern, air bronchogram sign, and bronchiolectasis. Moreover, bilateral and lower distribution of lung lesions in the periphery are defining characteristics of COVID-19, according to the disease severity and number of days [20, 20–23].

Since GGOs are common findings in lung images of COVID-19 patients [21, 22, 24], it is important to provide a proper definition. GGOs are hazy regions with slightly increased lung density and no shadowing of the bronchi and blood vessel margins. This can be attributed to the partial displacement of air caused by partial air filling or interstitial thickening [23, 25]. These patterns can be found In patients with COVID-19, one or multiple GGOs are typically found unilaterally or bilaterally in the subpleural region of the lung [22, 26, 27].

Relevant findings were determined with the following investigation on pregnant and perinatal women [28], which revealed that GGO was the most prevalent and earliest imaging feature, with an incidence rate ranging from 97% (81/83) to 100% [24, 29, 30]. The specific pathophysiological process underlying why GGOs are the earliest signs of CT is not clearly known [31, 32].

There are two computer-aided COVID CT diagnosis main methodologies and they are presented in the following subsections.

2.1.1 Image Level Diagnosis

First, this group of methods predicts the image level of a COVID-19-infected patient by providing a label to the original radiological image. To restrict the spread of the pandemic, these methods may be used for initial screening, preliminary analysis, and separation of infected individuals from healthy ones. For diagnosis and prognosis, AI models based on Deep Learning (DL) have been used to analyze medical images [33]. Image-level diagnostics assigns binary class labels (distinguishing COVID-19 infected images from Normal images) or multi-class labels (separating COVID-19 infected from Normal, viral Pneumonia, bacterial Pneumonia, etc.).

2.1.2 Region Level Diagnosis

In contrast to image-level diagnosis, region-level diagnosis predicts disease by labeling small patches or segmented sections inside a radiological picture. To define the region of interest, small areas of the image under analysis are categorized [33]. This allow future tracking of the infection and detection of the formation of different patterns.

Some investigations [1, 34] discovered that Ground-Glass Opacity (GGO) and other specific structures, such as consolidations, are usually found in COVID-19 lung images. The segmentation of GGO can be condensed into four major issues. (1) The shape and texture of GGO zones are highly changeable, making it difficult to identify fixed segmentation markers. (2) The poor contrast between GGO areas and edge tissues makes it difficult to precisely segment edges. (3) It is difficult to segment GGO edges using a single set threshold. (4) Value thresholds and their selection are issues that must be resolved. In addition, there are several threshold-based segmentation algorithms for COVID-19 GGO. It is challenging to modify the threshold and parameters, and the segmentation effect must be enhanced [35].

Medical images' segmentation is crucial to studying their pathological properties. Previously, segmentation was purely manual work, or sometimes, even only performing visual inspection and analysis every time the image was assessed. With the reduced cost and popularization of medical imaging examinations, computerized segmentation techniques becomes a valuable tool to make it possible for the specialist to propose the proper therapeutic protocols [15].

2.2 *COVID-19 CT Segmentation*

COVID-19-suspected patients require immediate diagnosis and treatment. CXR and CCT are routinely used to provide enough evidence for diagnosis. However, medical imaging, particularly chest CT scans, generates a large number of images, which makes diagnosis time-consuming for specialists. In addition, COVID-19 is a new disease with symptoms similar to other pneumonia types, necessitating the accumulation of extensive diagnostic experience by radiologists. Thus, AI-assisted medical image diagnosis is widely desired. The segmentation outlined in the preceding section might be used to preprocess the images. This chapter focuses on the techniques that could utilize the segmentation results in the diagnosis [36].

Currently, COVID-19 is one of the most dangerous infectious lung illnesses [17, 37]. Consequently, clinical imaging assessment is a frequent and essential procedure. Typically, the professional levels and subjective opinions of different physicians result in conflicting diagnostic outcomes [1, 38]. Although clinical medical technology possesses a set of quantitative indicators for preliminary screening [39–42], it is highly reliant on the gathering and quantification of image data, which hinders its promotion [43, 44]. Therefore, reconstruction of COVID-19 infection can give a digital model and quantitative analysis foundation for COVID-19 diagnosis and treatment. It contributes to the development of a more objective evaluation system and offers theoretical guidance to personnel in relevant domains [32, 45, 46].

As a result of the large number of lung infections caused by the COVID-19, the human population is facing a catastrophic and unprecedented worldwide health crisis. On CT scans, COVID-19 infection exhibits regional spread, fuzzy borders, tissue adhesion, and significant morphological changes. Therefore, establishing a method to effectively and accurately detect the infection from COVID-19 CT images is a significant and pressing task [15, 45].

CT scans have now become the most popular clinical imaging technique for COVID-19 assessment [3, 27, 47, 48] due to the fact that they can capture 3D patient information. Using thoracic CT as an example, the workflow for imaging-based COVID-19 diagnosis consists of three stages. The first stage is the "pre-scan preparation", with the setup of the tomography device and proper patient preparation. The second stage is the image acquisition itself, with specificities of this type of examination. The set of CT images is normally transferred to the PACS, where the medical team can have access to the third stage, which is the specialized diagnosis.

Segmentation is a crucial stage in image processing and analysis for COVID-19 evaluation and quantification. In chest X-ray or CT imaging, it outlines the relevant parts of the target organ or system, such as the lung, lobes, bronchopulmonary segments, and infected regions or lesions. These are called Regions of Interest (ROIs). The segmented regions could then be utilized to extract handcrafted or self-learned diagnostic characteristics for additional applications. This section would outline the segmentation-related contributions to COVID-19 and their applications. Widespread usage of deep learning approaches to segment ROIs in CT. Classic U-Net [8, 49–53], UNet++ [54, 55], and VB-Net [56] are well-known segmentation networks for COVID-19. Compared to CT, X-ray is more widely available worldwide. Segmenting X-ray images is sometimes more difficult by the projection of ribs onto soft tissues in 2D, which distorts image contrast. There is currently no developed method for segmenting X-ray pictures for COVID-19. However, Gaal et al. [57] use Attention-U-Net for lung segmentation in X-ray images for pneumonia. While the research is not specific to COVID-19, the method can be easily extended to the diagnosis of COVID-19 and other disorders.

The segmentation task is necessary for effective COVID-19 diagnosis [50, 53, 58, 59]. U-net has been successfully considered for this task in this study [51]. Other approaches are also used for medical imaging, such as quantification. The idea of quantification is to establish or propose metrics able to evaluate the severity or even the progression of the disease when comparing multiple examinations from the same patient. Several techniques and indexes are found in the literature [56, 60, 61]. For the COVID-19 lung images, multiple works present quantification techniques, such as longitudinal progression based on AI models [49], region-based segmentation for quantitative evaluation [8], and radiomic feature extraction of infected CT scans [52].

The segmentation approaches in COVID19 applications can be categorized primarily into two groups: lung-region-oriented methods and lung-lesion-oriented methods. The lung-region-oriented approaches attempt to separate lung areas, i.e., the total lung and lung lobes, from other (background) regions in CT or X-ray images, which is a prerequisite for COVID-19 applications. For instance, Jin et al.[55] suggest a two-stage pipeline for screening COVID-19 in CT images, in which an efficient network initially recognizes the entire lung region for segmentation based on UNet++ [54]. Due to the fact that the lung lesions or nodules present specific textures, forms, and sizes, their detection becomes a hard task to execute. For this issue, the inclusion of attention mechanisms in computerized learning and classification models may be an efficient localization strategy [57].

Recent attention techniques are able to discover the network's most discriminating features. Using an attention mechanism called "Attention U-Net", Oktay et al. [62] proposed a system that can detect small objects or lesions and are suitable to be used for COVID-19 medical images. A substantial amount of labeled data shall be necessary to train a robust segmentation AI model. Due to the labor-intensive and time-consuming manual delineation of lesions, sufficient training data for COVID-19 image segmentation tasks are usually unavailable. Incorporating human knowledge is a straightforward method for resolving this issue. For instance, Shan et al. [56] include radiologists in the training of a VB-net based segmentation network by including a human-in-the-loop technique. Qi et al. [52] define lung lesions using U-Net using radiologist-supplied starting seeds. Several further studies [60] using diagnostic knowledge and the attention mechanism found infection zones. When there are inadequate training data for segmentation, weakly-supervised machine learning techniques are also used. For example, Zheng et al. [53] propose constructing pseudo-segmentation masks for the pictures using an unsupervised approach, which is a trend in the area of medical imaging research, given the lack of annotated or labeled databases.

2.3 U-Net Network

U-Net is a convolutional neural network designed for biological picture segmentation at the University of Freiburg's Computer Science Department [7]. The basic assumption is to use Fully-Convolutional Networks (FCN) and the model has evolved to require fewer images for the training phase and improve segmentation performance. A contemporary GPU can segment a 512×512 picture in less than one second. Ronneberger [7] introduced the U-Net as a type of fully convolutional network with symmetric encoding and decoding signal channels with a U-shaped topology. The shortcut connections connect the same level's levels via two distinct paths. In this scenario, the network can gain superior visual semantics and contextures, making it suitable for segmenting medical images.

The FCNs were presented by Long, Shelhamer, and Darrell [5]. These networks aim to improve conventional deep learning models, adding successive layers with upsampling operators, instead of the conventional pooling operators. Therefore, these layers boost the output's resolution and a better classification performance can be achieved. In the upsampling portion of U-Net, a large number of feature channels allow the network to transmit context information to higher-resolution layers. As a result, the expanded path is roughly symmetrical with the contracting portion, resulting in a U-shaped architecture. The input images are mirrored in order to estimate the pixels in the image's border region. This tiling method is essential for applying the network to huge images since the GPU RAM would otherwise limit the resolution.

U-Net has numerous applications in biomedical imaging processing, especially for the segmentation task, including brain imaging [63], and even molecular analysis of proteins [64]. Medical image reconstruction has also utilized variants of the U-Net algorithm [65]. In [66], the U-Net was proposed with ImageNet-Trained VGG11 Encoder for Image Segmentation, while in [67] it was considered for Image-to-image translation for fluorescent stain estimation. Other applications of using the U-Net can be found in [68] and [69]. Several lung segmentation strategies [69–72] with diverse objectives have been published in the medical literature. In COVID applications, the U-Net method is widely utilized to segment lung areas and lung lesions.

A new segmentation model called UNet++ was proposed in the literature as a more complex evolution of the traditional U-Net model. This new model could improve segmentation performance and can be used to detect lung lesions. However, the training is more computationally demanding [73].

In addition, another work presented the ADID-UNET, which stands for Attention Gate—Dense Network—Improved Dilation Convolution—UNET. This model considers a dense network instead of the "convolution followed by maxpooling" traditional deep learning model. Using an enhanced dilation convolution, the receptive field of the encoder output is expanded in order to capture additional edge characteristics from the small infected patches. The new proposed model achieved both Accuracy and Specificity greater than 0.8, while the F1-score was 0.8031 and the Dice Coefficient was 0.82 [6].

A new model called EL-CNN-DF (Ensemble Learning—CNN-based Deep Features) is proposed by [74]. The method consists of a first phase of image filtering, followed by the segmentation step using U-net. The new technique is then applied to extract image features. Finally, a classification phase is implemented with three machine learning models: SVM—Support Vector Machines, Autoencoder classifier, and Naive Bayes (NB).

Another proposed model for lung lesions segmentation is the Content-Aware Residual UNet, also known as CARes-UNet. Considering the public dataset "COVID-SemiSeg", the proposed model achieved a Dice Coefficient of 0.776, outperforming other models [75].

Nguyen et al. integrated the Unet and Feature Pyramid Network (FPN) for COVID segmentation tasks using Computed Tomography (CT) scanner samples from the Italian Society of Medical and Interventional Radiology dataset. Experiments indicate that the decoder-based Unet family achieved its highest performance (a mean Intersection Over Union (mIoU) of 0.9234, 0.9032 in dice score, and a recall of 0.9349) with the combination of SE ResNeXt and Unet++. The decoder belonging to the Unet family achieved superior COVID segmentation performance than Feature Pyramid Network. In addition, the suggested method surpasses contemporary state-of-the-art segmentation techniques such as the SegNet-based network, ADID-UNET, and A-SegNet + FTL. Therefore, it is anticipated that it will deliver effective medical picture segmentation visualization [76].

2.4 Dataset

The dataset contains 349 CT scans that have been classified as positive for COVID-19. These CT pictures come in various sizes. Minimum, mean, and maximum height are 153, 491 and 1853 inches, respectively. Minimum, median, and maximum width are 124, 383, and 1485 respectively. These images come from 216 different patient instances. Among the 169 patients identified as positive, 137 had age information and 169 have gender information, with 86 male patients and 51 female patients [77] (Fig. 1).

2.5 Evaluation Criteria

This chapter experiment will consider DICE loss function as evaluation criteria. Dice Coefficient is a technique frequently used in digital libraries, the sciences, and other sectors to determine object similarity. This is the first attempt to use the Dice Coefficient for selecting conference papers [78]. An experimental result based on a small number of test cases shows that Dice Coefficient has the potential to be utilized across the entire spectrum of its relevant application. The definition of the Dice loss function DICE is as follows:

$$\text{dice\_loss} = 1 - \frac{1}{c}\sum_{i=0}^{c} \frac{\sum_j^N 2y_i^j \hat{y}_i^j + \epsilon}{\sum_j^N y_i^j + \sum_j^N \hat{y}_i^j + \epsilon}$$

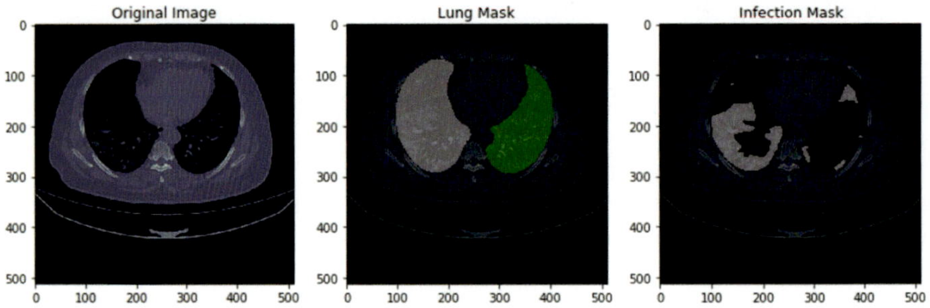

Fig. 1 Image sample from dataset

or utilizing squares in the denominator (DICE SQUARE), as suggested by Milletari1:

$$\text{dice\_loss\_square} = 1 - \frac{1}{c}\sum_{i=0}^{c} \frac{\sum_j^N 2y_i^j \hat{y}_i^j + \epsilon}{\sum_j^N y_i^j y_i^j + \sum_j^N \hat{y}_i^j \hat{y}_i^j + \epsilon}$$

ϵ is used to prevent division by 0 (denominator) and to learn from reference patches with no pixels of the (i)th class (nominator). The multiplication by $\frac{1}{c}$ has the desirable quality that, independent of the channel count, the loss lies within [0, 1]. Because it penalizes false positives, the dice loss can optionally be computed just for foreground channels (DICEFG, DICEFG SQUARE).

3 Experimental Results and Discussion

ADAM stochastic optimizer is used to train neural networks because of its rapid convergence rate compared to other optimizers. The input images are reduced to 100 versus 100 in order to reduce training time and resource needs. The image dataset is partitioned into three sets for training, validation, and testing, with relative proportions of 0.7, 0.2, and 0.1. In spite of class imbalance, median frequency balancing is used to produce class weights, which are then passed to the pixel classification layer to formulate a weighted cross-entropy loss function. After training the model with cross-entropy loss, we also employ a Dice loss-based fine-tuning strategy.

Figure 2 present a result obtained by the trained network. For the development of the U-Net network, the keras-unet framework with different configurations of filter parameters, normalization, dropout and number of layers in the network. The configuration that obtains the lowest cost for the loss function use normalization with four layers.

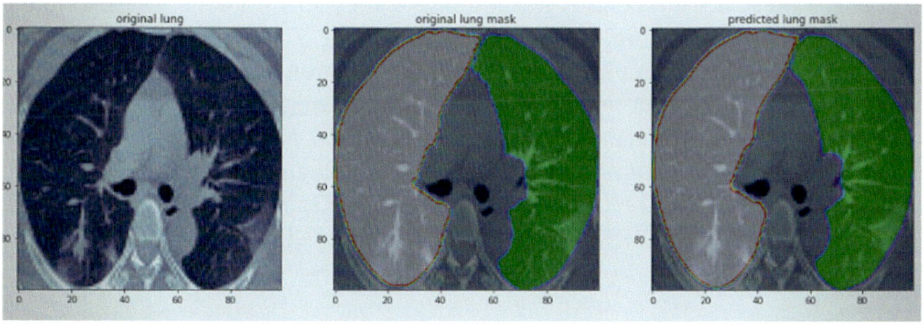

Fig. 2 Results obtained from dataset

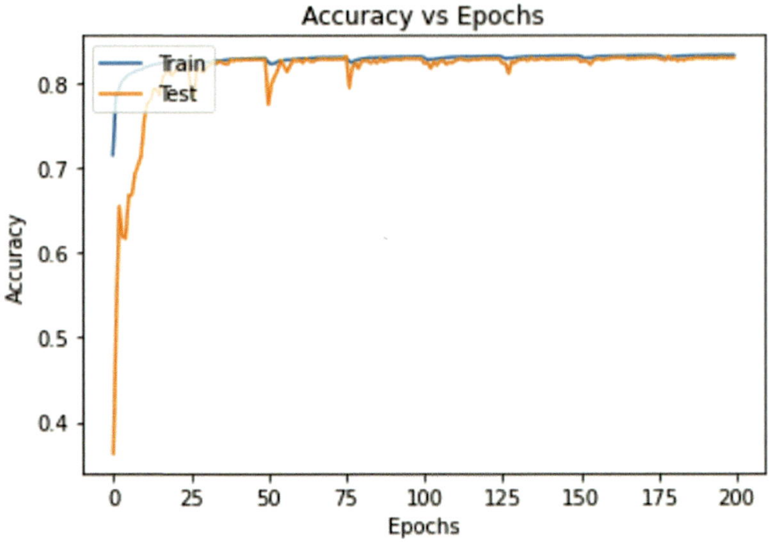

Fig. 3 Dice coefficient by epoch

Listing 1: Unet network

```
from keras_unet.models import custom_unet
input_shape = x[0].shape
model = custom_unet(input_shape,
    filters=32,
    use_batch_norm=True,
    dropout=0.3,
    dropout_change_per_layer=0.0,
    num_layers=4)
```

Figure 3 present the dice coefficient along 200 epochs training. The maximum value obtained was 0.83 (Fig. 4).

Figure 5 presents example of infection segmentation images obtained by the model. After segmentation we use the nibabel framework to perform a 3d reconstruction of the lung obtained (https://nipy.org/nibabel/). Nibabel provides access to read +/− write to some common medical and neuroimaging file formats, including: ANALYZE (simple, SPM99, SPM2 and later), GIFTI, NIfTI1, NIfTI2, CIFTI-2, MINC1, MINC2, AFNI BRIK/HEAD, MGH and ECAT, as well such as Philips PAR/REC.

Listing 2: 3D lung reconstruction with nibabel framework

```
import nibabel
converted_array = np.array(masks, dtype=np.float32)
converted_array=converted_array*255
affine = np.eye(4)
nifti_file = nibabel.Nifti1Image(converted_array, affine)
nibabel.save(nifti_file, 'dicom_segmentado.nii')
```

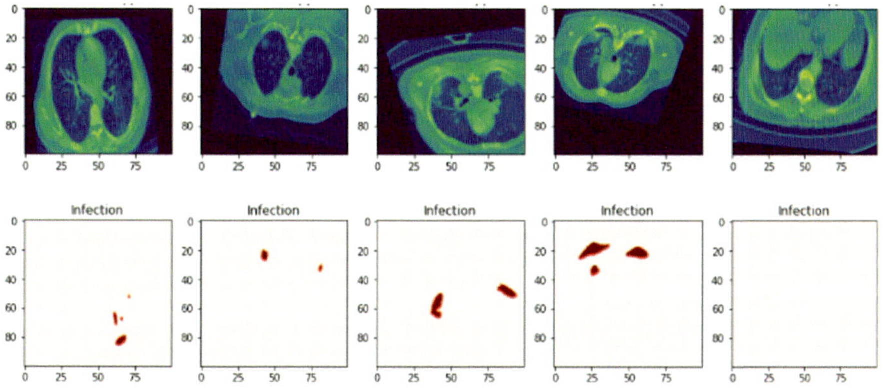

Fig. 4 Infection segmentation using training model

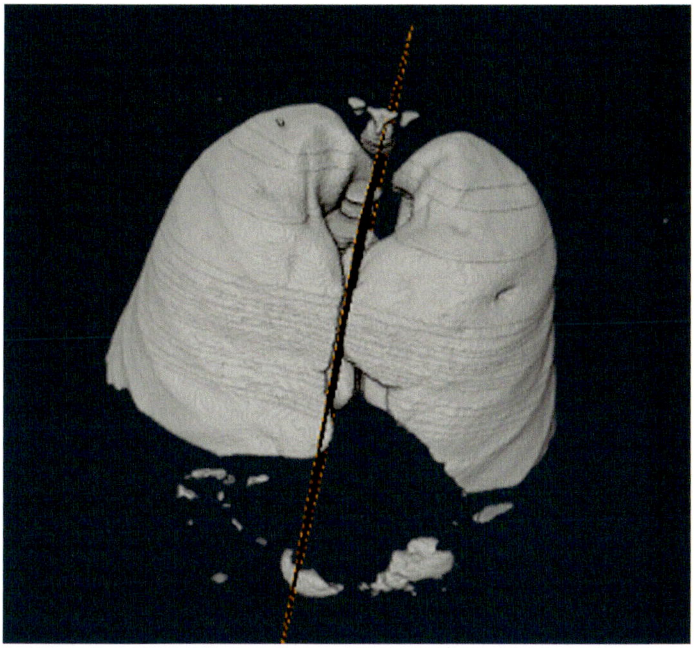

Fig. 5 3D reconstruction of the lung with Nibabel framework

4 Conclusion

Chest CT scan images are an efficient tool to assess COVID-19 lung symptoms, complementing or substituting laboratory tests, such as RT-PCR. CT scans show imaging characteristics that are comparable to other types of pneumonia, making it difficult to separate them. Lung infection manual representation is tedious and often influenced by personal bias and clinical competence.

Deep learning-based automatic lung segmentation of the COVID-19 patients examinations may support the specialist analysis to fastly assess the disease severity. Due to its efficacy and durability, U-Net is often used in medical picture segmentation. It cannot handle various features or account for each convolution channel's unique contribution to feature extraction. Convolution uses a local connection mode and weight sharing to minimize computation parameters, unlike the fully-connected layer dramatically. Local perception helps the convolution kernels gather info around pixels. This makes distant image pixels harder to relate to. The convolution kernel's local receptive field limits its utilization of image context information. Many prior efforts concentrated on enhancing segmentation at the pixel level, ignoring the importance of pixel grouping in semantic segmentation [17]. Our SD-UNet squeeze-and-attention (SA) module overcomes these challenges by using a not fully-squeezed attention channel mechanism to generate non-local spatial attention to the image and by using global context information to selectively re-weight channel features. It is also pixel grouping spatial awareness. Attention convolution scanned each pixel on the input feature map and grouped pixels with different spatial coordinates but the same class.

Ronneberger introduced the U-Net, a fully convolutional network with symmetric encoding and decoding signal channels and a U-shaped architecture. Two pathways connect the same level's shortcut connections. This situation gives the network better visual semantics and contextures for medical image segmentation. Diverse U-Net and its variants partition COVID-19 applications well. Long, Shelhamer, and Darrell introduced the "fully convolutional network" that underpins the U-Net design. Upsampling operators replace pooling operations in successive levels of a typical contracting network. These layers increase output resolution. This knowledge allows a later convolutional layer to assemble a precise segmentation output [7].

This chapter's experiment utilizes the DICE loss function as a criterion for evaluation. Dice Coefficient is a technique commonly used to determine object similarity in digital libraries, the sciences, and other fields. This is the first attempt to choose conference papers using the Dice Coefficient. We train a UNET network with the ADAM stochastic optimizer because of its rapid convergence rate compared to other optimizers. The number of input images is lowered to 100 versus 100 in order to reduce training time and resource requirements. The image dataset is divided into three sets with proportions of 70%, 20%, and 10% for training, validation, and testing, respectively. Despite the class imbalance, median frequency balancing is employed to generate class weights, which are then sent to the pixel classification layer to compute a weighted cross-entropy loss function. After training the model with cross-entropy loss, we also use a Dice loss-based technique for fine-tuning. Along 200 epochs training, we obtain the maximum value of 0.83 for the dice coefficient.

References

1. Ai T, Yang Z, Hou H, Zhan C, Chen C, Lv W, Tao Q, Sun Z, Xia L (2020) Correlation of chest ct and rt-pcr testing in coronavirus disease 2019 (covid-19) in china: a report of 1014 cases. Radiology
2. Fang Y, Zhang H, Xie J, Lin M, Ying L, Pang P, Ji W (2020) Sensitivity of chest ct for covid-19: comparison to rt-pcr. Radiology
3. Chung M, Bernheim A, Mei X, Zhang N, Huang M, Zeng X, Cui J, Xu W, Yang Y, Fayad ZA et al (2020) Ct imaging features of 2019 novel coronavirus (2019-ncov). Radiology
4. Schlemper J, Oktay O, Schaap M, Heinrich M, Kainz B, Glocker B, Rueckert D (2019) Attention gated networks: learning to leverage salient regions in medical images. Med Image Anal 53:197–207
5. Long J, Shelhamer E, Darrell T (2015) Fully convolutional networks for semantic segmentation. In: Proceedings of the IEEE conference on computer vision and pattern recognition, pp 3431–3440
6. Joseph Raj N, Zhu H, Khan A, Zhuang Z, Yang Z, Mahesh VGV, Karthik G (2021) Adid-unet-a segmentation model for covid-19 infection from lung ct scans. PeerJ Comput Sci 7:e349
7. Ronneberger O, Fischer P, Brox T (2015) U-net: convolutional networks for biomedical image segmentation. In: International conference on medical image computing and computer-assisted intervention. Springer, pp 234–241
8. Huang L, Han R, Ai T, Yu P, Kang H, Tao Q, Xia L (2020) Serial quantitative chest ct assessment of covid-19: a deep learning approach. Radiol Cardiothorac Imaging 2(2)
9. Shamim S, Javed Awan M, Mohd Zain A, Naseem U, Abed Mohammed M, Garcia-Zapirain B (2022) Automatic covid-19 lung infection segmentation through modified unet model. J Healthc Eng 2022
10. Wang S, Kang B, Ma J, Zeng X, Xiao M, Guo J, Cai M, Yang J, Li Y, Meng X et al (2021) A deep learning algorithm using ct images to screen for corona virus disease (covid-19). Eur Radiol 31(8):6096–6104
11. Rajaraman S, Siegelman J, Alderson PO, Folio LS, Folio LR, Antani SK (2020) Iteratively pruned deep learning ensembles for covid-19 detection in chest x-rays. Ieee Access 8:115041–115050
12. Zhou T, Canu S, Ruan S (2021) Automatic covid-19 ct segmentation using u-net integrated spatial and channel attention mechanism. Int J Imaging Syst Technol 31(1):16–27
13. Wang G, Liu X, Li C, Zhiyong X, Ruan J, Zhu H, Meng T, Li K, Huang N, Zhang S (2020) A noise-robust framework for automatic segmentation of covid-19 pneumonia lesions from ct images. IEEE Trans Med Imaging 39(8):2653–2663
14. Fan D-P, Zhou T, Ji G-P, Zhou Y, Chen G, Huazhu F, Shen J, Shao L (2020) Inf-net: automatic covid-19 lung infection segmentation from ct images. IEEE Trans Med Imaging 39(8):2626–2637
15. Yin S, Deng H, Zelin X, Zhu Q, Cheng J (2022) Sd-unet: a novel segmentation framework for ct images of lung infections. Electronics 11(1):130
16. Wu D, Gong K, Daniela Arru C, Homayounieh F, Bizzo B, Buch V, Ren H, Kim K, Neumark N, Xu P et al (2020) Severity and consolidation quantification of covid-19 from ct images using deep learning based on hybrid weak labels. IEEE J Biomed Health Inform 24(12):3529–3538
17. Dong E, Ratcliff J, Goyea TD, Katz A, Lau R, Ng TK, Garcia B, Bolt E, Prata S, Zhang D et al (2022) The johns hopkins university center for systems science and engineering covid-19 dashboard: data collection process, challenges faced, and lessons learned. In: The lancet infectious diseases

18. Bhattacharya S, Reddy Maddikunta PK, Pham Q-V, Reddy Gadekallu T, Lal Chowdhary C, Alazab M, Jalil Piran Md et al (2021) Deep learning and medical image processing for coronavirus (covid-19) pandemic: a survey. Sustain Cities Soc 65:102589
19. Chen N, Zhou M, Dong X, Jieming Q, Gong F, Han Y, Qiu Y, Wang J, Liu Y, Wei Y et al (2020) Epidemiological and clinical characteristics of 99 cases of 2019 novel coronavirus pneumonia in wuhan, china: a descriptive study. Lancet 395(10223):507–513
20. Bernheim A, Mei X, Huang M, Yang Y, Fayad ZA, Zhang N, Diao K, Lin B, Zhu X, Li K et al (2020) Chest ct findings in coronavirus disease-19 (covid-19): relationship to duration of infection. Radiology
21. Pan F, Ye T, Sun P, Gui S, Liang B, Li L, Zheng D, Wang J, Hesketh RL, Yang L et al (2020) Time course of lung changes at chest ct during recovery from coronavirus disease 2019 (covid-19). Radiology 295(3):715–721
22. Shi H, Han X, Jiang N, Cao Y, Alwalid O, Jin G, Fan Y, Zheng C (2020) Radiological findings from 81 patients with covid-19 pneumonia in wuhan, china: a descriptive study. Lancet Infect Dis 20(4):425–434
23. Zuo H (2020) Contribution of ct features in the diagnosis of covid-19. Can Respir J 2020
24. Zhou S, Wang Y, Zhu T, Xia L et al (2020) Ct features of coronavirus disease 2019 (covid-19) pneumonia in 62 patients in wuhan, china. Ajr Am J Roentgenol 214(6):1287–1294
25. Hansell DM, Bankier AA, MacMahon H, McLoud TC, Muller NL, Remy J et al (2008) Fleischner society: glossary of terms for thoracic imaging. Radiology 246(3):697
26. Pan Y, Guan H, Zhou S, Wang Y, Li Q, Zhu T, Qiongjie H, Xia L (2020) Initial ct findings and temporal changes in patients with the novel coronavirus pneumonia (2019-ncov): a study of 63 patients in wuhan, china. Eur Radiol 30(6):3306–3309
27. Song F, Shi N, Shan F, Zhang Z, Shen J, Lu H, Ling Y, Jiang Y, Shi Y (2020) Emerging 2019 novel coronavirus (2019-ncov) pneumonia. Radiology
28. Liu D, Li L, Wu X, Zheng D, Wang J, Liang B, Yang L, Zheng C (2020) Pregnancy and perinatal outcomes of women with covid-19 pneumonia: a preliminary analysis. Available at SSRN 3548758
29. Cheng Z, Yong L, Cao Q, Qin L, Pan Z, Yan F, Yang W (2020) Clinical features and chest ct manifestations of coronavirus disease 2019 (covid-19) in a single-center study in shanghai, china. Am J Roentgenol 215(1):121–126
30. Li K, Wu J, Wu F, Guo D, Chen L, Fang Z, Li C (2020) The clinical and chest ct features associated with severe and critical covid-19 pneumonia. Investig Radiol
31. Rui Han L, Huang HJ, Dong J, Peng H, Zhang D et al (2020) Early clinical and ct manifestations of coronavirus disease 2019 (covid-19) pneumonia. AJR Am J Roentgenol 215(2):338–43
32. Resende Lucinda Mangia L, Botelho Soares M, Sasso Carmona de Souza T, David João De Masi R, Cristina Scarabotto P, Hamerschmidt R (2021) Objective evaluation and predictive value of olfactory dysfunction among patients hospitalized with covid-19. Auris Nasus Larynx 48(4):770–776
33. Çallı E, Sogancioglu E, van Ginneken B, van Leeuwen KG, Murphy K (2021) Deep learning for chest x-ray analysis: a survey. Med Image Anal 72:102125
34. Ye Z, Zhang Y, Wang Y, Huang Z, Song B (2020) Chest ct manifestations of new coronavirus disease 2019 (covid-19): a pictorial review. Eur Radiol 30(8):4381–4389
35. Wai Lee K, Ka Yin Chin R (2021) An adaptive data processing framework for cost-effective covid-19 and pneumonia detection. In: 2021 IEEE international conference on signal and image processing applications (ICSIPA). IEEE, pp 150–155
36. Shi F, Wang J, Shi J, Ziyan W, Wang Q, Tang Z, He K, Shi Y, Shen D (2020) Review of artificial intelligence techniques in imaging data acquisition, segmentation, and diagnosis for covid-19. IEEE Rev Biomed Eng 14:4–15
37. Bilal Tahir M, Batool A (2020) Covid-19: healthy environmental impact for public safety and menaces oil market. Sci Total Environ 740:140054
38. Jajodia A, Ebner L, Heidinger B, Chaturvedi A, Prosch H (2020) Imaging in corona virus disease 2019 (covid-19)-a scoping review. Eur J Radiol Open 7:100237

39. Dang Y, Liu N, Tan C, Feng Y, Yuan X, Fan D, Peng Y, Jin R, Guo Y, Lou J (2020) Comparison of qualitative and quantitative analyses of covid-19 clinical samples. Clinica Chimica Acta 510:613–616
40. Lanza E, Muglia R, Bolengo I, Giuseppe Santonocito O, Lisi C, Angelotti G, Morandini P, Savevski V, Salvatore Politi L, Balzarini L (2020) Quantitative chest ct analysis in covid-19 to predict the need for oxygenation support and intubation. Eur Radiol 30(12):6770–6778
41. Mori M, Palumbo D, De Lorenzo R, Broggi S, Compagnone N, Guazzarotti G, Giorgio Esposito P, Mazzilli A, Steidler S, Pietro Vitali G et al (2021) Robust prediction of mortality of covid-19 patients based on quantitative, operator-independent, lung ct densitometry. Physica Medica 85:63–71
42. Yoon H, O'Neill Byerley C, Joshua S, Moore K, Sook Park M, Musgrave S, Valaas L, Drimalla J (2021) United states and south korean citizens' interpretation and assessment of covid-19 quantitative data. J Math Behav 62:100865
43. Al-Antari MA, Hua C-H, Bang J, Lee S (2021) Fast deep learning computer-aided diagnosis of covid-19 based on digital chest x-ray images. Appl Intell 51(5):2890–2907
44. Saygılı A (2021) A new approach for computer-aided detection of coronavirus (covid-19) from ct and x-ray images using machine learning methods. Appl Soft Comput 105:107323
45. Chen C, Zhou J, Zhou K, Wang Z, Xiao R (2021) Dw-unet: loss balance under local-patch for 3d infection segmentation from covid-19 ct images. Diagnostics 11(11):1942
46. Larenas-Linnemann D, Rodríguez-Pérez N, Antonio Ortega-Martell J, Blandon-Vijil V, Luna-Pech JA (2020) Coronavirus disease 2019 and allergen immunotherapy: theoretical benefits invite to adjustments in practice recommendations. Ann Allergy Asthma Immunol 125(3):247–249
47. Cheng Z, Qin L, Cao Q, Dai J, Pan A, Yang W, Gao Y, Chen L, Yan F (2020) Quantitative computed tomography of the coronavirus disease 2019 (covid-19) pneumonia. Radiol Infect Dis 7(2):55–61
48. Ng M-Y, Lee EYP, Yang J, Yang F, Li X, Wang H, Mei-sze Lui M, Shing-Yen Lo C, Leung B, Khong P-L et al (2020) Imaging profile of the covid-19 infection: radiologic findings and literature review. Radiol Cardiothorac Imaging 2(1)
49. Cao Y, Xu Z, Feng J, Jin C, Han X, Wu H, Shi H (2020) Longitudinal assessment of covid-19 using a deep learning–based quantitative ct pipeline: illustration of two cases. Radiol Cardiothorac Imaging 2(2)
50. Gozes O, Frid-Adar M, Greenspan H, Browning PD, Zhang H, Ji W, Bernheim A, Siegel E (2020) Rapid ai development cycle for the coronavirus (covid-19) pandemic: Initial results for automated detection & patient monitoring using deep learning ct image analysis. arXiv:2003.05037
51. Li L, Qin L, Xu Z, Yin Y, Wang X, Kong B, Bai J, Lu Y, Fang Z, Song Q et al (2020) Artificial intelligence distinguishes covid-19 from community acquired pneumonia on chest ct. Radiology
52. Qi X, Jiang Z, Yu Q, Shao C, Zhang H, Yue H, Ma B, Wang Y, Liu C, Meng X et al (2020) Machine learning-based ct radiomics model for predicting hospital stay in patients with pneumonia associated with sars-cov-2 infection: a multicenter study. MedRxiv
53. Zheng C, Deng X, Fu Q, Zhou Q, Feng J, Ma H, Liu W, Wang X (2020) Deep learning-based detection for covid-19 from chest ct using weak label. MedRxiv
54. Chen J, Lianlian W, Zhang J, Zhang L, Gong D, Zhao Y, Chen Q, Huang S, Yang M, Yang X et al (2020) Deep learning-based model for detecting 2019 novel coronavirus pneumonia on high-resolution computed tomography. Sci Rep 10(1):1–11
55. Jin S, Wang B, Xu H, Luo C, Wei L, Zhao W, Hou X, Ma W, Xu Z, Zheng Z et al (2020) Ai-assisted ct imaging analysis for covid-19 screening: building and deploying a medical ai system in four weeks. MedRxiv
56. Shan F, Gao Y, Wang J, Shi W, Shi N, Han M, Xue Z, Shen D, Shi Y (2020) Lung infection quantification of covid-19 in ct images with deep learning. arXiv:2003.04655
57. Gaál G, Maga B, Lukács A (2020) Attention u-net based adversarial architectures for chest x-ray lung segmentation. arXiv:2003.10304

58. Jin C, Chen W, Cao Y, Zhanwei X, Tan Z, Zhang X, Deng L, Zheng C, Zhou J, Shi H et al (2020) Development and evaluation of an artificial intelligence system for covid-19 diagnosis. Nat Commun 11(1):1–14
59. Song Y, Zheng S, Li L, Zhang X, Zhang X, Huang Z, Chen J, Wang R, Zhao H, Chong Y et al (2021) Deep learning enables accurate diagnosis of novel coronavirus (covid-19) with ct images. IEEE/ACM Trans Comput Biol Bioinform 18(6):2775–2780
60. Shen C, Nan Yu, Cai S, Zhou J, Sheng J, Liu K, Zhou H, Guo Y, Niu G (2020) Quantitative computed tomography analysis for stratifying the severity of coronavirus disease 2019. J Pharm Anal 10(2):123–129
61. Tang L, Zhang X, Wang Y, Zeng X (2020) Severe covid-19 pneumonia: assessing inflammation burden with volume-rendered chest ct. Radiol Cardiothorac Imaging 2(2)
62. Oktay O, Schlemper J, Le Folgoc L, Lee M, Heinrich M, Misawa K, Mori K, McDonagh S, Hammerla NY, Kainz B et al (2018) Attention u-net: learning where to look for the pancreas. arXiv:1804.03999
63. Baid U, Ghodasara S, Mohan S, Bilello M, Calabrese E, Colak E, Farahani K, Kalpathy-Cramer J, Kitamura FC, Pati S et al (2021) The rsna-asnr-miccai brats 2021 benchmark on brain tumor segmentation and radiogenomic classification. arXiv:2107.02314
64. Nazem F, Ghasemi F, Fassihi A, Mehri Dehnavi A (2021) 3d u-net: a voxel-based method in binding site prediction of protein structure. J Bioinform Comput Biol 19(02):2150006
65. Andersson J, Ahlström H, Kullberg J (2019) Separation of water and fat signal in whole-body gradient echo scans using convolutional neural networks. Magn Reson Med 82(3):1177–1186
66. Iglovikov V, Shvets A (2018) Ternausnet: U-net with vgg11 encoder pre-trained on imagenet for image segmentation. arXiv:1801.05746
67. Kandel ME, He YR, Jae Lee Y, Hsuan-Yu Chen T, Michele Sullivan K, Aydin O, Saif MTA, Kong H, Sobh N, Popescu G (2020) Phase imaging with computational specificity (pics) for measuring dry mass changes in sub-cellular compartments. Nat Commun 11(1):6256
68. Yao W, Zeng Z, Lian C, Tang H (2018) Pixel-wise regression using u-net and its application on pansharpening. Neurocomputing 312:364–371
69. Çiçek Ö, Abdulkadir A, Lienkamp SS, Brox T, Ronneberger O (2016) 3d u-net: learning dense volumetric segmentation from sparse annotation. In: Medical image computing and computer-assisted intervention–MICCAI 2016: 19th international conference, Athens, Greece, October 17-21, 2016, proceedings, part II 19. Springer, pp 424–432
70. Isensee F, Petersen J, Klein A, Zimmerer D, Jaeger PF, Kohl S, Wasserthal J, Koehler G, Norajitra T, Wirkert S et al (2018) nnu-net: self-adapting framework for u-net-based medical image segmentation. arXiv:1809.10486
71. Milletari F, Navab N, Ahmadi S-A (2016) V-net: Fully convolutional neural networks for volumetric medical image segmentation. In: 2016 fourth international conference on 3D vision (3DV). IEEE, pp 565–571
72. Zhou Z, Siddiquee MMR, Tajbakhsh N, UNet+ Liang J (2018) A nested u-net architecture for medical image segmentation. arXiv:1807.10165
73. Zhou Z, Mahfuzur Rahman Siddiquee Md, Tajbakhsh N, Liang J (2018) Unet++: a nested u-net architecture for medical image segmentation. In: Deep learning in medical image analysis and multimodal learning for clinical decision support. Springer, pp 3–11
74. Das A (2022) Adaptive unet-based lung segmentation and ensemble learning with cnn-based deep features for automated covid-19 diagnosis. Multimed Tools Appl 81(4):5407–5441
75. Xu X, Wen Y, Zhao L, Zhang Y, Zhao Y, Tang Z, Yang Z, Yu-Chian Chen C (2021) Cares-unet: content-aware residual unet for lesion segmentation of covid-19 from chest ct images. Med Phys 48(11):7127–7140
76. Thanh Nguyen H, Bao Tran T, Hoang Luong H, Khoi Nguyen Huynh T (2021) Decoders configurations based on unet family and feature pyramid network for covid-19 segmentation on ct images. PeerJ Comput Sci 7:e719
77. Covid-19 ct scans | kaggle (2023). https://www.kaggle.com/datasets/andrewmvd/covid19-ct-scans. (Accessed on 01/19/2023)
78. Nurhilyana A, Md Sultan AB (2010) Validate conference paper using dice coefficient. Comput Inf Sci 3(3):139

Covid-19 Detection Based on Chest X-Ray Images Using Multiple Transfer Learning CNN Models

Bruno Riccelli dos Santos Silva, Paulo Cesar Cortez, Pedro Crosara Motta, and Joao Alexandre Lobo Marques

Abstract The gold standard to detect SARS-CoV-2 infection considers testing methods based on Polymerase Chain Reaction (PCR). Still, the time necessary to confirm patient infection can be lengthy, and the process is expensive. In parallel, X-Ray and CT scans play an important role in the diagnosis and treatment processes. Hence, a trusted automated technique for identifying and quantifying the infected lung regions would be advantageous. Chest X-rays are two-dimensional images of the patient's chest and provide lung morphological information and other characteristics, like ground-glass opacities (GGO), horizontal linear opacities, or consolidations, which are typical characteristics of pneumonia caused by COVID-19. This chapter presents an AI-based system using multiple Transfer Learning models for COVID-19 classification using Chest X-Rays. In our experimental design, all the classifiers demonstrated satisfactory accuracy, precision, recall, and specificity performance. On the one hand, the Mobilenet architecture outperformed the other CNNs, achieving excellent results for the evaluated metrics. On the other hand, Squeezenet presented a regular result in terms of recall. In medical diagnosis, false negatives can be particularly harmful because a false negative can lead to patients being incorrectly diagnosed as healthy. These results suggest that our Deep Learning classifiers can accurately classify X-ray exams as normal or indicative of COVID-19 with high confidence.

B. R. dos Santos Silva · P. Cesar Cortez · P. Crosara Motta
Laboratório de Engenharia e Sistemas de Computação, Universidade Federal do Ceará, Campus do Pici, Fortaleza, Brazil
e-mail: cortez@lesc.ufc.br

P. Crosara Motta
e-mail: pedro.motta@lesc.ufc.br

J. A. Lobo Marques (✉)
Laboratory of Applied Neurosciences, University of Saint Joseph, Estrada Marginal da Ilha Verde, 14-17, Macao SAR, China
e-mail: alexandre.lobo@usj.edu.mo

J. A. Lobo Marques and S. J. Fong (eds.), *Computerized Systems for Diagnosis and Treatment of COVID-19*, https://doi.org/10.1007/978-3-031-30788-1_4

1 Introduction

The diagnosis of COVID-19 using CXR (Chest X-Rays) is challenging for both radiology specialists and computerized diagnostic supporting systems. When compared to similar CT Scans, the image quality is poorer, and the potential capacity of analysis is highly lower since the tomography provides hundreds of slices of the lungs under study. Nevertheless, CT scanners are much more expensive equipment, and the examination procedure is complex and costly. In contrast, X-Ray examinations are simple to execute and widely available in most clinical facilities.

In some cases, it is possible to easily detect changes in the lung conditions caused by COVID-19 pneumonia by assessing an X-Ray image. In other cases, X-Ray images are poor and noisy and demand filtering and other computational techniques to extract features in order to classify the presence of infection.

Given the exponential need for medical imaging examinations for COVID-19 patients and the lack of radiology specialists to provide reporting, the use of Artificial Intelligence (AI) tools to analyze these medical images and provide diagnostic support for the medical teams has been considered. More recently, Deep Learning models are achieving significant results for medical imaging classification, segmentation, and clustering, among other tasks, considering multiple metrics, such as Accuracy, Sensitivity, and Specificity.

Deep Learning models are commonly based on Convolutional Neural Networks (CNN). A complex task usually demands a complex model to achieve satisfactory performance, which results in the need for expensive GPU farms with high-processing power and large amounts of memory to train and validate the model. This is not readily available for any software development team, and pre-trained models validated for specific tasks are considered to be reused for different applications with some level of tuning for the new context to be considered, which is known as Transfer Learning and will be explained in detail in the following Section.

The main objective of this chapter is twofold. First, a general overview of different Transfer Learning approaches is presented to comprehend the possibilities of creating a classification from previously developed models. Second, to compare the classification performance for a public database of Chest X-Ray examinations of COVID-19 and normal patients.

2 Transfer Learning

Transfer learning is a machine learning method where a model developed for a task is reused as the starting point for a model on a second task. It is an increasingly popular strategy for applying deep learning approaches considering pre-trained models (architecture and weights) to be adopted as the starting point for achieving one task. Currently, the applications are mainly in the areas of image processing/computer vision and natural language processing. These applications demand heavy computing power (often with expensive GPU farms) for a long time, creating a burden of resources (computational and financial).

Several more comprehensive definitions are provided in the literature. Goodfellow, Bengio, and Courville state that Transfer Learning and domain adaptation refer to the situation where what has been learned in one setting is exploited to improve generalization in another setting [1]. From another perspective, Goldberg and Hirst define Transfer learning as the improvement of learning in a new task through the transfer of knowledge from a related task that has already been learned [16].

Many possibilities of Transfer Learning models can be found in the literature and some key references are provided as follows:

- VGG-16 and VGG-19 [13]
- GoogleNet [14]
- AlexNet [7]
- ResNet-18, ResNet-50, ResNet-101 [3]
- SqueezeNet [6]
- MobileNet-V2 [11].

A general diagram representing the process of Transfer Learning, reusing the model proposed for Problem A for the necessary classification in Problem B, is presented in Fig. 1.

The process is inductive, i.e. from the empirical model developed for Problem A, a new model is obtained for Problem B. There are two significant risks: the possibility of reusing a biased model, and, even if the model is not biased, it may have poor generalization capacity to execute different related tasks.

For implementing Transfer Learning properly, there are two common model approaches:

- Develop Model Approach - is considered when data is easily available for Problem A, then a model is implemented to solve that specific task. The resultant model is then reused and adjusted to solve Problem B, with a proper phase for adjustments and model tunning.
- Pre-trained Model Approach - this is when a model is selected from existing available libraries. During the last few years, several research institutions have released challenging data for the academic community and made reusable models available. This became a widespread approach not only for model definition but also for model validation.

Three popular groups of Transfer Learning models available are from Oxford, known as VGG Models; from Google, called Inception Models; and from Microsoft, the ResNet Models. In this chapter, the VGG16 model is considered, and the application is explained further.

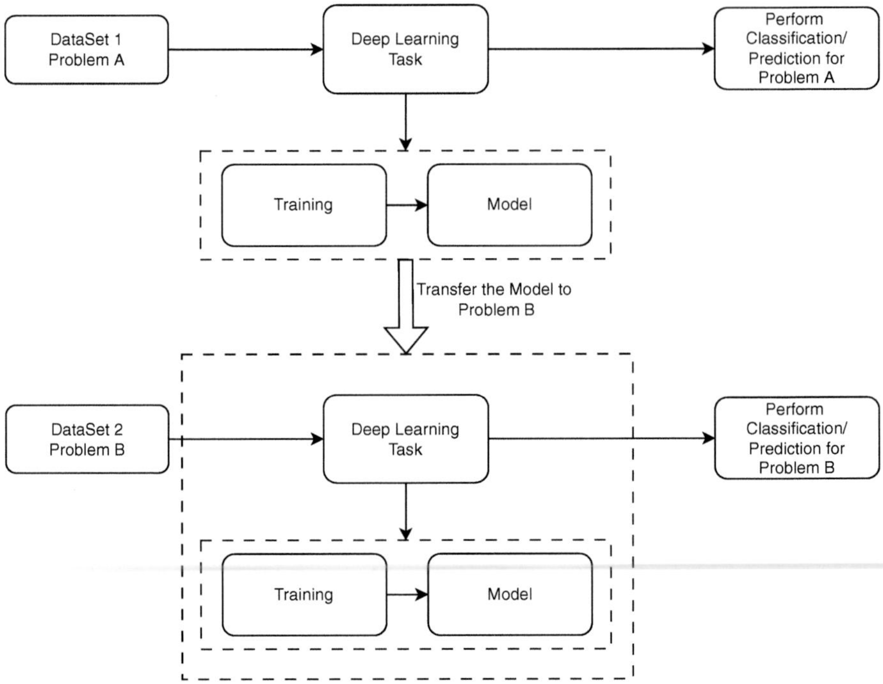

Fig. 1 General diagram representing the process of Transfer Learning, reusing the model proposed for Problem A for the necessary classification in Problem B

3 Experimental Methodology

The methodology used in this study is a comparison between state-of-the-art CNN architectures for Covid-19 classification. The COVIDx-CXR2 dataset is a collection of chest X-ray images that have been labeled as either normal or COVID-19 infection. Researchers at the University of Oxford and the University of Hong Kong created the dataset. The dataset contains over 30,000 images, with a balanced distribution of normal and COVID-19 exams divided in train and test folders. We separate 30% of the training set for the validation one.

3.1 Proposed Experiment

In this chapter, we propose a comprehensive comparison between eight CNNs applied to Covid-19 classification from X-ray images. Figure 2 illustrates the proposed methodology.

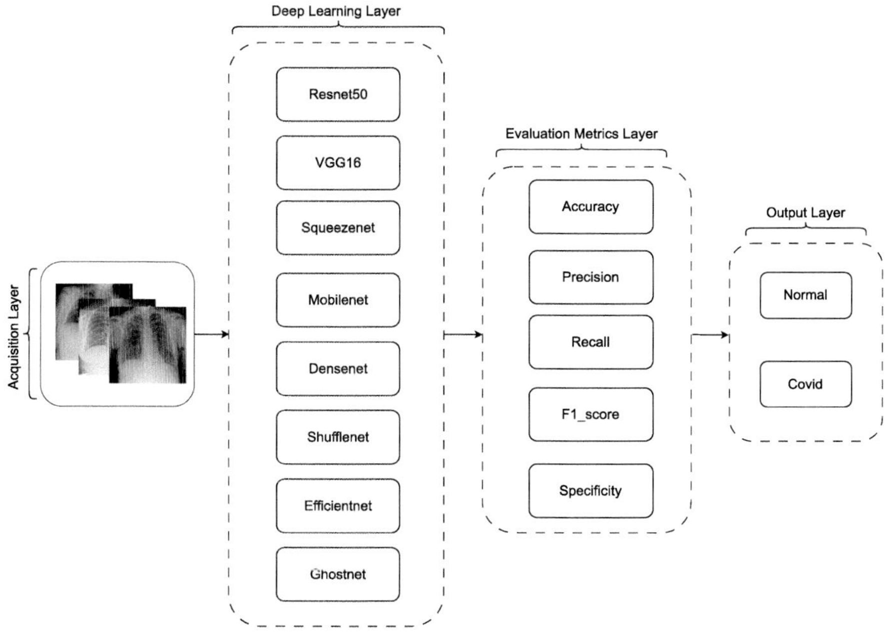

Fig. 2 Methodology flowchart for this chapter

There are four layers in the proposed pipeline. The acquisition layer is responsible for the chest x-ray reception. The Deep Learning layer comprises the eight CNNs applied in this work, Resnet50, VGG16, SqueezeNet, MobileNet, DenseNet, ShuffleNet, EfficientNet, and Ghostnet. These are state-of-the-art classification methodologies that will be evaluated for Covid classification. The classifiers are evaluated through Accuracy, Precision, Recall, F1 and Specificity. Finally, the output layer will be responsible for getting the classification of the best metric and thus classifying the exam in normal or covid. We applied 30 epochs at training phase and for preventing overffiting, we used early stopping technique, monitoring the validation accuracy. Thus, if there is no improvement for this metric in more than five epochs, the training phase terminates.

3.2 *Performance Evaluation*

In this chapter, we evaluated the classifiers using Accuracy, F1-Score, Precision, Recall and Specificity. These formulations are described as:

- Accuracy:

$$Acc = \frac{TP + TN}{TP + TN + FP + FN} \tag{1}$$

- Precision:

$$P = \frac{TP}{TP + FP} \tag{2}$$

- Recall :

$$R = \frac{TP}{TP + FN} \tag{3}$$

- F1-Score:

$$F1 = \frac{2TP}{2TP + FP + FN} \tag{4}$$

- Specificity:

$$S = \frac{TN}{TN + FP} \tag{5}$$

Accuracy is the proportion between correct predictions and all predictions. Is the simplest and most common evaluation metric, but can be lead to miss interpretations when the classes are imbalanced. Precision is the proportion of correct positive predictions. In the one hand, it can be used in applications where False positives are considered more harmful than False negatives [9]. On the other hand, Recall represents the fraction of positive cases, so it can be applied when the false negative rate is crucial. F1-score is the harmonic mean of precision and recall. Specificity is the proportion of true negative predictions among all negative ones.

4 Experiments

In our experiment design,we used eight different Deep Learning classifiers to classify X-ray exams as either normal or indicative of COVID-19. The architecture explanations and obtained results are presented in the following sections.

4.1 Resnet50

ResNet50 is a convolutional neural network architecture proposed by *He et al.* [3]. It is composed by 50 layers that are based on the residual concept. The main concept behind residual connections is that the network learns the difference, or residual, between the input and output rather than the entire mapping. This allows the network to more easily learn the identity function and can be trained deeper, preventing the vanishing gradient problem. Figure 3 illustrates the Resnet50 architecture.

First, the zero padding layer preserves the spatial dimensions of the input before the convolutional layer. It prevents information loss at the edges and controls the output size. The initial CNN and max pooling layers reduce the dimensions of the

Fig. 3 Resnet50 architecture

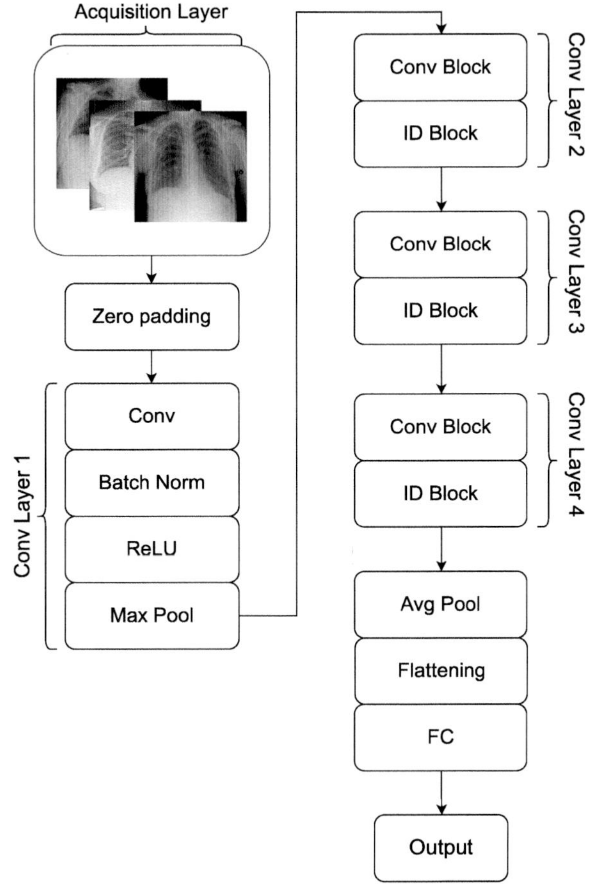

input image and increase the number of channels. The residual units are represented by convolutional layers, batch normalization and ReLu activations. Moreover there is a global average pooling layer that is applied to reduce spatial dimensions of feature maps to a single value per channel. In the end of the architecture, it have a fully connected layer with a single output to make the final prediction.

The Resnet50 has been pre-trained on the ImageNet dataset, a large dataset of images for classification task. This network is largely utilized for tasks such biomedical imaging [10], object detection [12] and Intrusion Detection Systems [8].

Table 1 presents validation and test results achieved by Resnet50 architecture.

Overall the model performed well in validation and testing sets, achieving high values of accuracy, F1, Precision, Recall and Specificity. This indicates that this model is capable to identify both classes correctly. The achieved precision was 100%, meaning that of all normal exams were classified correctly. However, the recall of 90.66% indicates that is still possibilities of improvement, because 9.34% of the

Table 1 Validation and test results obtained for Resnet50 architecture

	Validation (%)	Test (%)
Accuracy	98.63	95.62
F1-score	98.71	95.10
Precision	98.97	100
Recall	98.45	90.66
Specificity	98.83	100

Covid-19 exams were classified as normal. Note that the recall metric is related to false negative as previously mentioned.

4.2 VGG-16

VGG-16 is a convolutional neural network architecture proposed by *Symonian et al.* [13]. Figure 4 presents the VGG-16 architecture.

The architecture of VGG-16 is composed of 23 layers, which are divided into 5 blocks. Each block consists of 3 or 4 convolutional layers, followed by a max pooling layer. The number of filters in each convolutional layer increases with each block. Additionally, the VGG-16 architecture uses only 3×3 convolutional filters, which helps to reduce the number of parameters.

As the Resnet50 architecture, the VGG-16 is trained on the ImageNet dataset, obtaining a good performance and still widely used for computer vision task such as semantic segmentation

Table 2 presents validation and test results achieved by VGG-16 architecture.

Overall the model performed very well in validation and testing sets, achieving high values of accuracy, F1, Precision, Recall and Specificity. This indicates that this model is capable to identify both classes correctly. Moreover, only 4% of the Covid-19 exams were classified as normal.

4.3 Squeezenet

Iandola et al. [6] presented a lightweight deep convolutional neural network designed for low cost hardware requirements called SqueezeNet. Figure 5 presents the Squeezenet architecture.

The fire blocks are building blocks that consists of a squeeze layer and an expand layer. The squeeze layer reduces the number of feature maps and the expand layer increases it. This allows for a more efficient use of computation resources, as well as reducing the number of parameters in the model.

Table 3 presents validation and test results achieved by Squeezenet architecture.

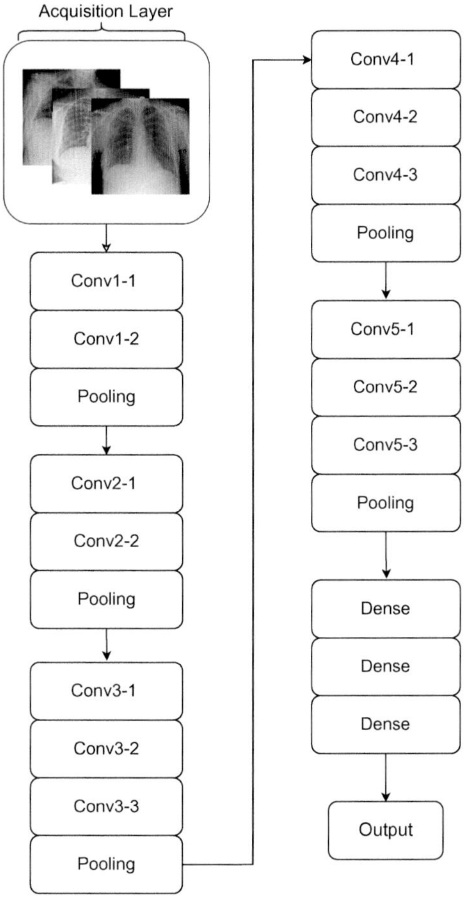

Fig. 4 VGG-16 architecture

Table 2 Validation and test results obtained for VGG-16 architecture

	Validation %	Test %
Acc	98.62	97.50
F1	98.71	97.29
Precision	98.18	98.63
Recall	99.24	96.00
Specificity	97.90	98.82

Overall the model performed well in validation set, achieving high values of accuracy, F1, Precision, Recall and Specificity. This indicates that this model is capable to identify both classes correctly in this stage. However, the recall of 82.66% indicates that is still possibilities of improvement, because 17.34% of the Covid-19 exams were classified as normal. Note that the recall metric is related to false negative as previously mentioned.

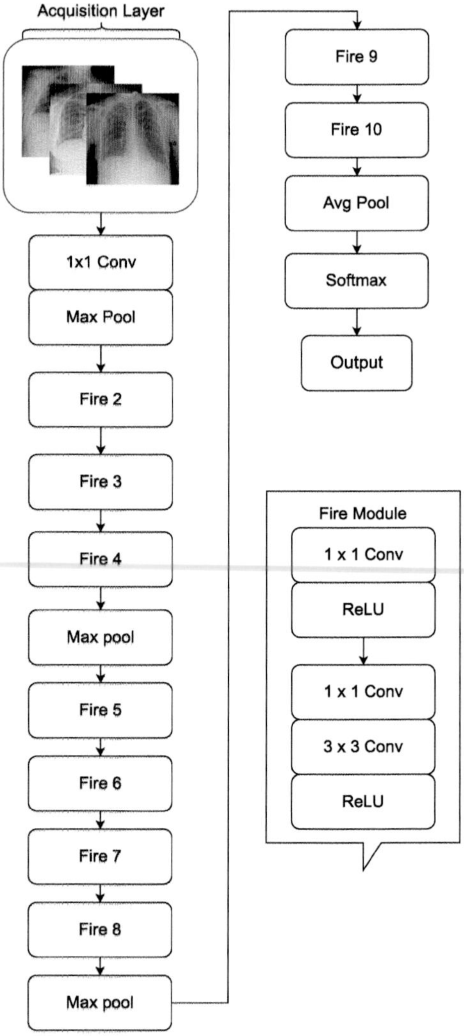

Fig. 5 Squeezenet architecture

Table 3 Validation and test results obtained for Squeezenet architecture

	Validation %	Test %
Acc	97.92	91.25
F1	98.03	89.85
Precision	98.54	98.41
Recall	97.53	82.66
Specificity	98.36	98.82

4.4 Mobilenet

Another lightweight architecture is the Mobilenet. It was developed by *Howard et al.* [4]. Figure 6 illustrates its architecture.

An important feature of MobileNet is the use of the MobileNetV2 architecture, which include an inverted residual block with a linear bottleneck, this architecture allows to reduce the computational cost while increasing the representational power of the model.

Table 4 presents validation and test results achieved by Mobilenet architecture.

Overall the model performed well in validation and testing sets, achieving high values of accuracy, F1, Precision, Recall and Specificity. This indicates that this

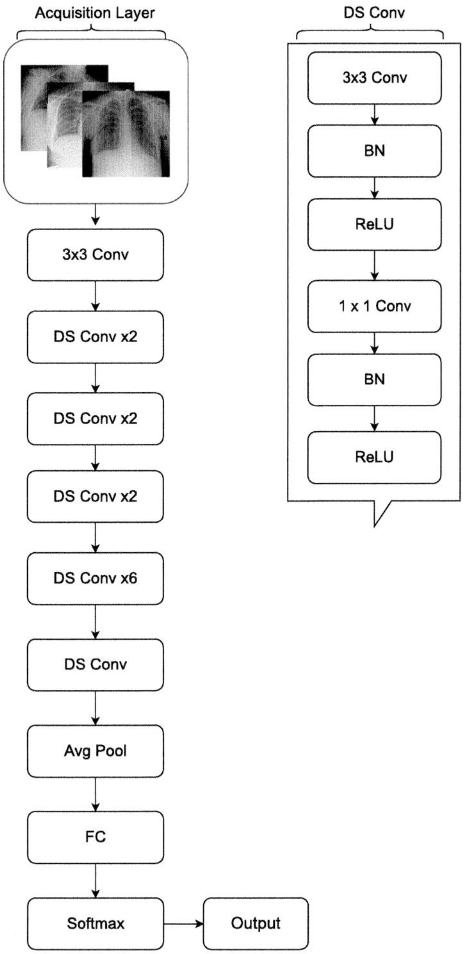

Fig. 6 Mobilenet architecture

Table 4 Validation and test results obtained for Mobilenet architecture

	Validation %	Test %
Acc	98.49	98.75
F1	98.59	98.64
Precision	98.15	100
Recall	99.03	97.33
Specificity	97.88	100

model is capable to identify both classes correctly. The achieved precision was 100%, meaning that of all normal exams were classified correctly. Moreover, only 2.67% of the Covid-19 exams were classified as normal.

4.5 Densenet-201

Gao Huang *et al.* [5] proposed a CNN called Densenet-201. It is an extension of ResNet using denses connections between the layers. Figure 7 illustrates the Densenet-201 architecture.

The input image in acquisition layer is passed through convolutional layers, for feature extraction. The features are passed through the dense blocks and combined with features from other layers in the network. The output of each dense block is then passed through a transition layer, which reduces the spatial resolution of the feature maps.

Table 5 presents validation and test results achieved by Densenet architecture.

Overall the model performed well in validation and testing sets, achieving high values of accuracy, F1, Precision, Recall and Specificity. This indicates that this model is capable to identify both classes correctly. The achieved precision was 100%, meaning that of all normal exams were classified correctly. However, the recall of 92% indicates that is still possibilities of improvement, because 8% of the Covid-19 exams were classified as normal. Note that the recall metric is related to false negative as previously mentioned.

4.6 Shufflenet

Another Lightweight network was proposed by *Zhang et al.* [17] aiming efficient computation on mobile devices. Some techniques were combined in its construction such as point-wise group convolution, channel shuffle and a bottleneck structure. Figure 8 illustrates the Shufflenet architecture.

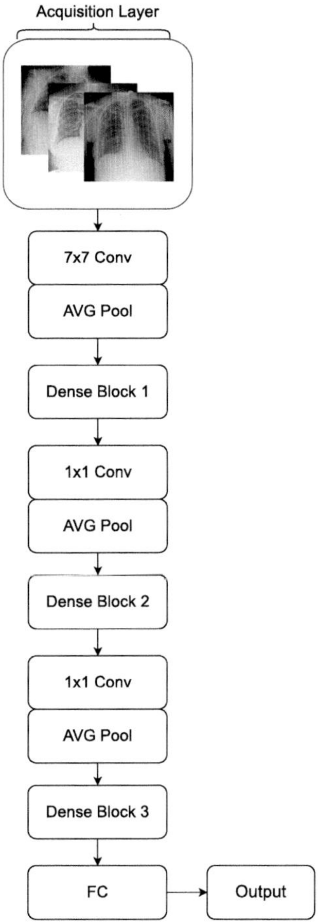

Fig. 7 Densenet architecture

Table 5 Validation and test results obtained for Densenet architecture

	Validation %	Test %
Acc	98.63	96.24
F1	98.70	95.83
Precision	99.23	100
Recall	98.18	92.00
Specificity	99.14	100

The ShuffleNet architecture uses pooling and concatenation layers to reduce the feature map sizes and increase the number of channels of the feature maps, respectively. It is composed of several stages, consisting of multiple ShuffleUnits and one bottleneck structure. The output of each stage is concatenated with the output of the previous stage.

Table 6 presents validation and test results achieved by Shufflenet architecture.

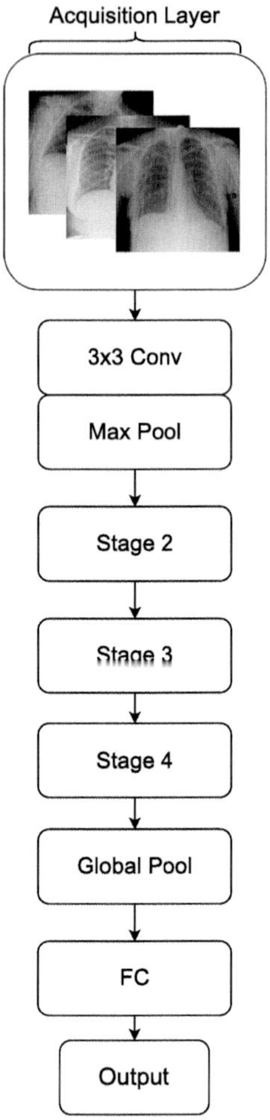

Fig. 8 Shufflenet architecture

Table 6 Validation and test results obtained for Shufflenet architecture

	Validation %	Test %
Acc	98.37	96.24
F1	98.47	95.83
Precision	98.41	100
Recall	98.53	92.00
Specificity	98.19	100

Overall the model performed well in validation and testing sets, achieving high values of accuracy, F1, Precision, Recall and Specificity. This indicates that this model is capable to identify both classes correctly. The achieved precision was 100%, meaning that of all normal exams were classified correctly. However, the recall of

92% indicates that is still possibilities of improvement, because 8% of the Covid-19 exams were classified as normal. Note that the recall metric is related to false negative as previously mentioned.

4.7 *Efficientnet*

It is another lightweight network was proposed by *Tan et al.* [15]. EfficientNet is an improvement of MobileNet Architecture scaling up the model in a more efficient way adjusting depth, width, and resolution of the model simultaneously. Figure 9 illustrates the Efficientnet architecture.

The EfficientNet architecture utilizes the MBConv block, which combines a depthwise convolution, a pointwise convolution, and a squeeze-and-excitation operation. The depthwise convolution increases the number of channels in the feature maps, while the pointwise convolution reduces it. The squeeze-and-excitation operation adapts the feature maps with global information.

Table 7 presents validation and test results achieved by Shufflenet architecture.

Overall the model performed well in validation and testing sets, achieving high values of accuracy, F1, Precision, Recall and Specificity. This indicates that this model is capable to identify both classes correctly. The achieved precision was 100%, meaning that of all normal exams were classified correctly. However, the recall of 85.33% indicates that is still possibilities of improvement, because 14.67% of the Covid-19 exams were classified as normal. Note that the recall metric is related to false negative as previously mentioned.

4.8 *Ghostnet*

GhostNet is a recent lightweight CNN presented by*Han et al.* [2] that was developed for edge devices, focusing mainly in object detection and classification tasks with low computational complexity. GhostNet uses a combination of depthwise separable convolutions, which reduces the number of parameters, and a unique Ghost module that enhances feature representation. Figure 10 illustrates the Ghostnet architecture.

The Ghost module uses a combination of a squeeze-and-excitation operation and an attention mechanism to re-calibrate the feature maps by using global information.

Table 8 presents validation and test results achieved by Ghostnet architecture.

Overall the model performed well in validation and testing sets, achieving high values of accuracy, F1, Precision, Recall, and Specificity. This indicates that this model is capable of identifying both classes correctly. The achieved precision was 100%, meaning that all normal exams were classified correctly. However, the recall of 90.66% indicates that is still a possibility of improvement because 9.34% of the Covid-19 exams were classified as normal.

Fig. 9 Efficientnet architecture

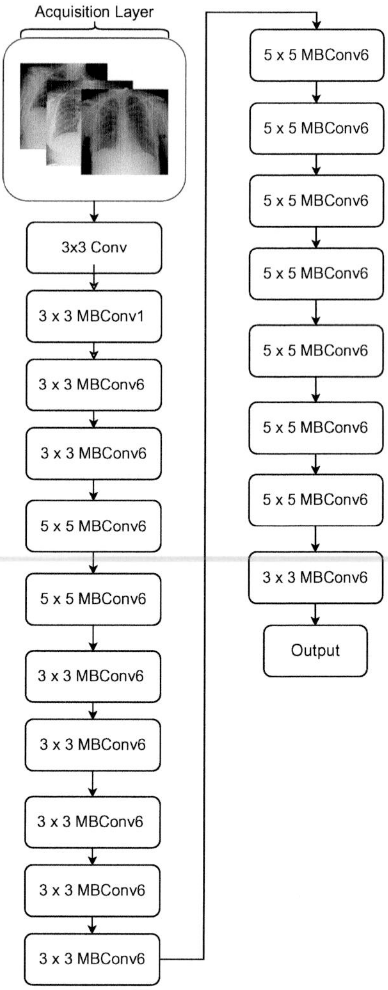

Table 7 Validation and test results obtained for efficientnet architecture

	Validation %	Test %
Acc	98.14	93.12
F1	98.25	92.08
Precision	98.54	100
Recall	97.95	85.33
Specificity	98.36	100

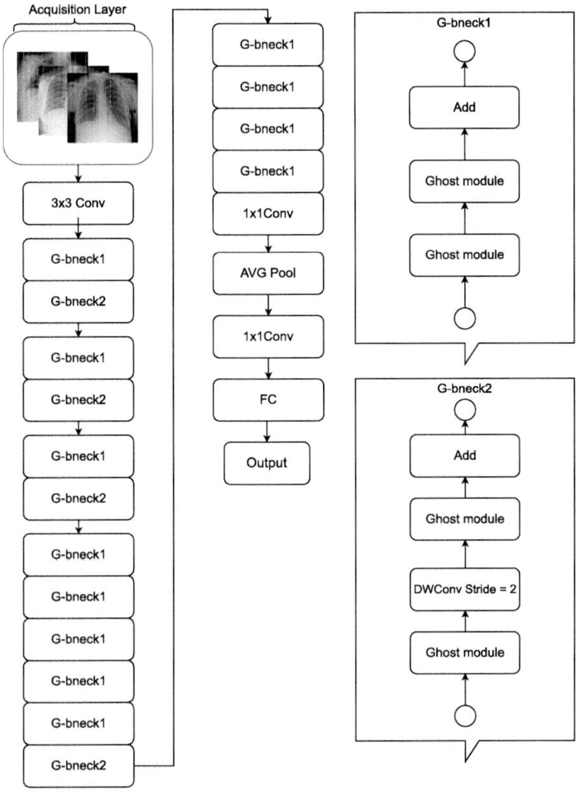

Fig. 10 Ghostnet architecture

Table 8 Validation and test results obtained for Ghostnet architecture

	Validation %	Test %
Acc	98.15	95.62
F1	98.25	95.10
Precision	99.12	100
Recall	97.38	90.66
Specificity	99.02	100

5 Consolidated Results

In this section, we present and discuss the obtained results of this chapter. Table 9 illustrates the metrics achieved by all CNNs in the testing phase with 400 x-ray images not trained by the models.

In our experimental design, all the classifiers demonstrated satisfactory accuracy, precision, recall, and specificity performance. On the one hand, the Mobilenet architecture outperformed the other CNNs, achieving excellent results for the evaluated metrics. On the other hand, Squeezenet presented a regular result in terms of recall.

Table 9 Consolidated results of this chapter

Architecture	Accuracy %	F1 %	Precision %	Recall %	Specificity %
Resnet50	95.62	95.10	100	90.66	100
VGG-16	97.50	97.29	98.63	96.00	98.82
Squeezenet	91.25	89.85	98.41	82.66	98.82
Mobilenet	98.75	98.64	100	97.33	100
Densenet	96.24	95.83	100	92.00	100
Shufflenet	96.24	95.83	100	92.00	100
Efficientnet	93.12	92.08	100	85.33	100
Ghostnet	95.62	95.10	100	90.66	100

The best Accuracy was achieved by the Mobilenet with 98.75% with the lowest by the Squeezenet with 91.24%. Six models achieved a Precision of 100%, Resnet50, Mobilenet, Densenet, Shufflenet, Efficientnet, and Ghostnet, while VGG-16 and Squeezenet achieved 98.63% and 98.41%, respectively. The best Recall was achieved by the Mobilenet with 97.33% with the lowest by the Squeezenet with 82.66%. For Specificity, Six models achieved 100%, Resnet50, Mobilenet, Densenet, Shufflenet, Efficientnet, and Ghostnet, while VGG-16 and Squeezenet achieved both 98.82%. Finally, for the F1-score, the best performance again was achieved by the Mobilenet with 98.64% with the lowest by the Squeezenet with 89.85%.

In medical diagnosis, false negatives can be particularly harmful because a false negative can lead to patients being incorrectly diagnosed as healthy. These results suggest that our Deep Learning classifiers can accurately classify Chest X-Ray exams as normal or infected with COVID-19 with a promising confidence.

6 Conclusion

This chapter presented an AI-based system using multiple Transfer Learning models for COVID-19 classification using Chest X-Rays. In the proposed experimental design, different classifiers demonstrated satisfactory accuracy, precision, recall, and specificity performance. On the one hand, the Mobilenet architecture outperformed the other CNNs, achieving excellent results for the evaluated metrics. On the other hand, Squeezenet presented a regular result in terms of recall.

The best Accuracy was achieved by the Mobilenet with 98.75%. Six models (Resnet50, Mobilenet, Densenet, Shufflenet, Efficientnet, and Ghostnet) achieved 100% for Precision and Specificity. The best Recall was achieved by the Mobilenet with 97.33%. Finally, for the F1-score, the best performance was achieved by the Mobilenet model with 98.64%.

In medical diagnosis, false negatives can be particularly harmful because a false negative can lead to patients being incorrectly diagnosed as healthy. These results suggest that our Deep Learning classifiers can accurately classify Chest X-Ray exams as normal or infected with COVID-19 with promising confidence.

References

1. Goodfellow I, Bengio Y, Courville A (2016) Deep learning. MIT Press. http://www.deeplearningbook.org
2. Han K, Wang Y, Tian Q, Guo J, Xu C, Xu C (2020) Ghostnet: more features from cheap operations. In: Proceedings of the IEEE/CVF conference on computer vision and pattern recognition, pp 1580–1589
3. He K, Zhang X, Ren S, Sun J (2016) Deep residual learning for image recognition. In: Proceedings of the IEEE conference on computer vision and pattern recognition, pp 770–778
4. Howard AG, Zhu M, Chen B, Kalenichenko D, Wang W, Weyand T, Andreetto M, Adam H (2017) Mobilenets: efficient convolutional neural networks for mobile vision applications. arXiv:1704.04861
5. Huang G, Liu Z, Van Der Maaten L, Weinberger KQ (2017) Densely connected convolutional networks. In: Proceedings of the IEEE conference on computer vision and pattern recognition, pp 4700–4708
6. Iandola FN, Han S, Moskewicz MW, Ashraf K, Dally WJ, Keutzer K (2016) Squeezenet: alexnet-level accuracy with 50x fewer parameters and< 0.5 mb model size. arXiv:1602.07360
7. Krizhevsky A, Sutskever I, Hinton GE (2017) Imagenet classification with deep convolutional neural networks. Commun ACM 60(6):84–90
8. Rezende E, Ruppert G, Carvalho T, Ramos F, De Geus P (2017) Malicious software classification using transfer learning of resnet-50 deep neural network. In: 2017 16th IEEE international conference on machine learning and applications (ICMLA). IEEE, pp 1011–1014
9. Riccelli Silva B, Jardel Silveira R, Gonçalves da Silva Neto M, César Cortez P, Gonçalves Gomes D (2021) A comparative analysis of undersampling techniques for network intrusion detection systems design. J Commun Inf Syst 36(1):31–43
10. Sai Bharadwaj Reddy A, Sujitha Juliet D (2019) Transfer learning with resnet-50 for malaria cell-image classification. In: 2019 international conference on communication and signal processing (ICCSP). IEEE, pp 0945–0949
11. Sandler M, Howard A, Zhu M, Zhmoginov A, Chen L-C (2018) Mobilenetv2: inverted residuals and linear bottlenecks. In: Proceedings of the IEEE conference on computer vision and pattern recognition, pp 4510–4520
12. Shao S, Li Z, Zhang T, Peng C, Yu G, Zhang X, Li J, Sun J (2019) Objects365: a large-scale, high-quality dataset for object detection. In: Proceedings of the IEEE/CVF international conference on computer vision, pp 8430–8439
13. Simonyan K, Zisserman A (2014) Very deep convolutional networks for large-scale image recognition. arXiv:1409.1556
14. Szegedy C, Liu W, Jia Y, Sermanet P, Reed S, Anguelov D, Erhan D, Vanhoucke V, Rabinovich A (2015) Going deeper with convolutions. In: Proceedings of the IEEE conference on computer vision and pattern recognition, pp 1–9
15. Tan M, Le Q (2019) Efficientnet: rethinking model scaling for convolutional neural networks. In: International conference on machine learning. PMLR, pp 6105–6114
16. Torrey LA, Shavlik JW (2009) Transfer learning. MUniversity of Wisconsin
17. Zhang X, Zhou X, Lin M, Sun J (2018) Shufflenet: an extremely efficient convolutional neural network for mobile devices. In: Proceedings of the IEEE conference on computer vision and pattern recognition, pp 6848–6856

X-Ray Machine Learning Classification with VGG-16 for Feature Extraction

Bruno Riccelli dos Santos Silva, Paulo Cesar Cortez, Manuel Gonçalves da Silva Neto, and Joao Alexandre Lobo Marques

Abstract The Covid-19 pandemic evidenced the need Computer Aided Diagnostic Systems to analyze medical images, such as CT and MRI scans and X-rays, to assist specialists in disease diagnosis. CAD systems have been shown to be effective at detecting COVID-19 in chest X-ray and CT images, with some studies reporting high levels of accuracy and sensitivity. Moreover, it can also detect some diseases in patients who may not have symptoms, preventing the spread of the virus. There are some types of CAD systems, such as Machine and Deep Learning-based and Transfer learning-based. This chapter proposes a pipeline for feature extraction and classification of Covid-19 in X-ray images using transfer learning for feature extraction with VGG-16 CNN and machine learning classifiers. Five classifiers were evaluated: Accuracy, Specificity, Sensitivity, Geometric mean, and Area under the curve. The SVM Classifier presented the best performance metrics for Covid-19 classification, achieving 90% accuracy, 97.5% of Specificity, 82.5% of Sensitivity, 89.6% of Geometric mean, and 90% for the AUC metric. On the other hand, the Nearest Centroid (NC) classifier presented poor sensitivity and geometric mean results, achieving 33.9% and 54.07%, respectively.

B. R. dos Santos Silva · P. C. Cortez
Laboratório de Engenharia e Sistemas de Computação, Universidade Federal do Ceará, Campus do Pici, Fortaleza, Brazil
e-mail: bruno@lesc.ufc.br

P. C. Cortez
e-mail: cortez@lesc.ufc.br

M. G. da Silva Neto
Federal Institute of Piauí, Rua Antonio Martins de Andrade, 750, Pedro II Piauí, Brazil
e-mail: manuel@ifpi.edu.br

J. A. Lobo Marques (✉)
Laboratory of Applied Neurosciences, University of Saint Joseph, Estrada Marginal da Ilha Verde, 14-17, Macao SAR, China
e-mail: alexandre.lobo@usj.edu.mo

J. A. Lobo Marques and S. J. Fong (eds.), *Computerized Systems for Diagnosis and Treatment of COVID-19*, https://doi.org/10.1007/978-3-031-30788-1_5

1 Introduction

The Covid-19 pandemic evidenced the need for computerized solutions to support medical staff and public health systems [12]. Computer Aided Diagnosis to analyze medical images such as CT and MRI scans and X-rays to assist doctors in disease diagnosis. CAD systems have been shown to be effective at detecting COVID-19 in chest X-ray and CT images, with some studies reporting high levels of accuracy and sensitivity. Moreover, it can also detect some diseases in patients who may not have symptoms, preventing the spread of the virus. There are some types of CAD systems, such as Machine and Deep Learning-based and Transfer learning-based.

Transfer learning is a Machine learning field that focuses on the problem of adapting models trained on one task to perform well on a different task. The main idea is to leverage the knowledge learned from one task to improve the performance of a model on a second task by using the learned features and knowledge from the source task to the target one.

Transfer learning is particularly useful when limited data is available for the new task. Because the model has already learned from a related task, it can often learn the new task with much less data. It makes transfer learning a powerful technique for many practical applications such as computer vision, natural language processing, and speech recognition.

One particular technique of transfer learning is the VGG-16 architecture. It comprises 16 layers, including 13 convolutional layers and three fully connected layers, allowing it to learn fine-grained features from images. The usage of VGG-16 as a feature extractor comprises the remotion of the fully connected layers to get the convolutional layer output.

The literature presents works that used hand-engineered features for COVID-19 classification from X-ray images. These works usually used traditional machine learning algorithms rather than deep learning. In addition, the features used in these works are often based on image processing techniques, such as thresholding, morphological operations, and texture analysis.

One example of a paper that uses hand-engineered features for COVID-19 classification from X-ray images is the one proposed by Hussain et al. [10]. In this paper, the authors extract various features from the X-ray images and use them as inputs to five machine learning algorithms for classification through sensitivity, specificity, and AUC metrics. The authors ranked the extracted features to analyze their importance for distinguishing Covid-19 cases of normal, bacterial infection, and viral pneumonia.

Another example is the paper proposed by Ohata et al. [15]. In this paper, the authors used transfer learning techniques such as VGG-16, Resnet50, MobileNet, InceptionV3, and Xception for feature extraction. Then these features were applied to Support vector machines (SVM), kNN, Multilayer perceptron (MLP), Random forest, and Naive Bayes classifiers. Finally, they evaluated the classifiers using Accuracy, F1 score, False positive rate, training and testing time.

In this chapter, it is proposed a pipeline for feature extraction and classification of Covid-19 in X-ray images using techniques based on multiple approaches, such as distance, trees, and gradient.

2 Experimental Model

This section presents the materials and methods used in the experiments, describing the overall procedures that compose the evaluation scenarios.

2.1 Machine Learning for Biomedical Imaging

Applying machine learning to biomedical imaging involves utilizing algorithms and statistical models to analyze and interpret medical images, including X-rays, MRI, and CT scans. The objective is to enhance the precision and effectiveness of diagnosis, disease management, and treatment planning [13, 14].

The three main application areas of machine learning for biomedical imaging are clinical exam segmentation, disease classification, and Computer-Aided Diagnosis Systems. The first allows for identifying and separating organs or structures in an image for further analysis. The second involves detecting and categorizing different diseases or conditions based on image exams. The last one is applied for making accurate diagnoses by providing additional information and highlighting potential issues for supporting doctors.

There are several Machine learning classifiers employed for disease detection in the literature. In the proposed pipeline, it will be considered:

- Nearest Centroid (NC);
- K-Nearest Neighbors (kNN);
- Support Vector Machines (SVM);
- Random Forest (RF);
- Histogram of Gradient Boosting (HGB).

2.2 Proposed Methodology

Figure 1 presents a schema for model evaluation that measures the efficacy of the features extracted from VGG-16. In the proposed system, we employ the feature-based dataset extracted from the VGG-16 with well-established machine-learning algorithms to discriminate pathological cases from healthy ones. According to a data-centric processing flow, we organized the experiment design as follows. First, a feature-based dataset was extracted from the VGG-16 (see Sect. 2.2.2). Second, we

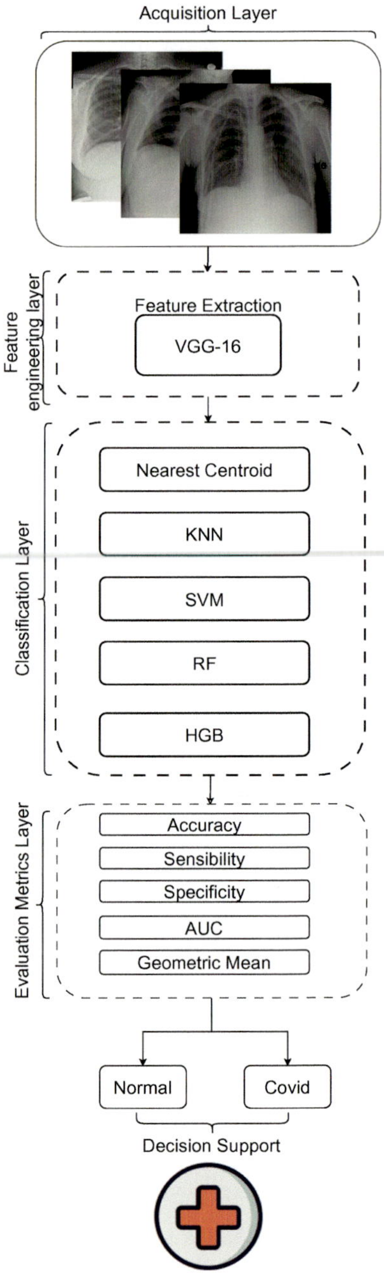

Fig. 1 Methodology flowchart for this chapter

employed feature-based machine learning algorithms for classification tasks (See Sect. 2.2.4). Third, the models were evaluated with metrics for binary classification in the biomedical domain (See Sect. 2.2.3). With the extracted features, the resultant dataset within the proposed procedure provided a schema to support the decision for machine learning models for Covid-19 detection from X-ray images.

2.2.1 Dataset

The COVIDx-CXR2 dataset [16] is one of the most widely used datasets for COVID-19 diagnosis from CXR images and has been used in several studies and publications in the field [1, 3, 4, 11, 19]. It is a collection of chest X-ray images labeled as either normal or COVID-19 infection. Researchers at the University of Oxford and the University of Hong Kong created the dataset. The dataset contains over 30,000 images, with a balanced distribution of normal and COVID-19 exams divided into train and test folders. In addition, we separate 30% of the training set for the validation one.

2.2.2 Feature Extraction

As previously mentioned, VGG-16 is a CNN widely exploited in image classification tasks. There are some steps that we can follow to perform a feature extraction using this CNN:

- Obtain a pre-trained VGG-16 model and load it into our workspace. To do this, we must choose the dataset used for pre-training the network. Some datasets are commonly used for pre-training purposes, such as ImageNet, COCO (Common Objects in Context), Pascal VOC (Visual Object Classes), and MS-COCO (Microsoft Common Objects in Context). In this chapter, we used the ImageNet dataset for the pre-training phase and got its weights to proceed with the feature extraction.
- After that, we remove some of the layers from the pre-trained VGG-16 model. We removed the fully connected layers and kept the convolutional layers. The convolutional layers' activations are our features.
- We passed our X-ray images through the modified VGG-16 model to extract the features from the activations of the intermediate layers.
- Finally, the output of VGG convolutional layers have a size of (1,4096). In this phase, the data have a shape of $(N, 1{,}4096)$, where N is the number of X-ray samples collected. To avoid this issue, we reshaped the data for two-dimensional data resulting in a $(N, 4096)$ shape.

It is worth mentioning that some approaches for feature engineering use transfer learning, such as Resnet50, Mobilenet, DenseNet, GhostNet, and SqueezeNet. These networks can provide competitive results in the biomedical area. Figure 2 presents the architecture of VGG-16 without classification layers.

Fig. 2 VGG-16 architecture

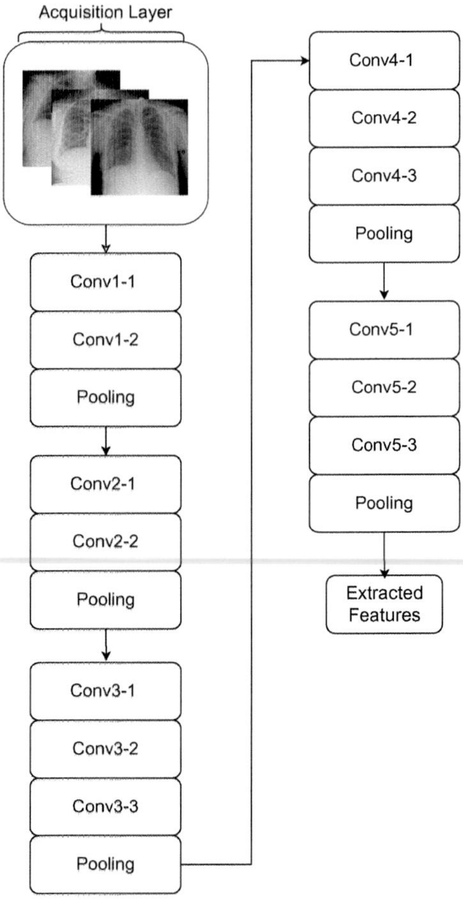

The VGG architecture have five convolution layers with max pooling layers, which helps reduce the spatial dimensions of the feature maps while retaining the most important information. Aiming to extract features, the three fully connected layers are removed and we can access the features.

2.2.3 Performance Metrics

The experimental design made use of evaluation metrics for biomedical analysis such as Accuracy (ACC), Specificity (SP), Sensitivity (SE), the area under the receiver operating characteristic curve (AUC), and the geometric mean (Gmean) between SP and SE. Let TP be the True positive rate, TN be the True negative rate, FP is the False positive rate, and FN is the False negative rate. The metric formulations are described as follows.

- Accuracy (ACC):

$$Acc = \frac{TP + TN}{TP + TN + FP + FN} \quad (1)$$

- Specificity (SP):

$$Sp = \frac{TN}{TN + FP} \quad (2)$$

- Sensitivity (SE) :

$$Se = \frac{TP}{TP + FN} \quad (3)$$

- Geometric mean (Gmean):

$$Gmean = \sqrt{SP * SE} \quad (4)$$

The metrics mentioned above are often used in studies to assess the performance of methods for machine learning algorithms [8, 18].

2.2.4 Feature-Based Models Evaluation

In conjunction with inner cross-validations for hyper-parameter search, the Holdout technique was successfully employed to evaluate feature-based machine learning models in the literature [9].

Figure 3 presents the model evaluation procedure. The Holdout technique was used to split the dataset into training and validation portions. Our experiments employed ten repetitions of the holdout to evaluate the classification algorithms. On each holdout repetition, an inner five-fold stratified cross-validation was employed on the training data to search the optimal classifier hyper-parameters. We also scaled the training data by applying the min-max scaling and applied the resulting scale to the validation data. Next, we used the optimal-parameter classifier to compute the evaluation metrics on the validation data. Finally, we averaged the holdout result metrics.

3 Experimental Results

In this section, we present and discuss the results obtained from the proposed pipeline.

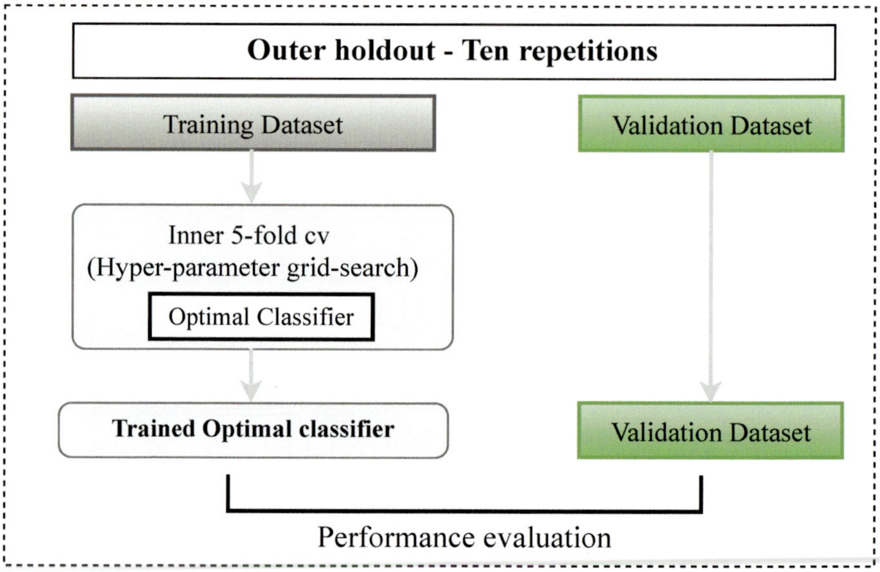

Fig. 3 Evaluation of the models

3.1 *Nearest Centroid (NC)*

The Nearest Centroid [17] algorithm was chosen as the baseline model in our experiments. It is a simple and fast machine-learning algorithm used for classification problems. It uses instance-based learning, where each class is represented by its members' average (centroid). The Nearest Centroid Classifier is simple to implement and efficient, especially when dealing with high-dimensional data. However, it is sensitive to the scale of the features and may not perform well when there is a lot of noise in the data.

Table 1 illustrates the Accuracy, Specificity, Sensibility, Gmean, and AUC with their respective mean and standard deviation (Std) for the NC classifier.

The Nearest Centroid classifier presented poor performance metrics, achieving sensitivity of 33.9%. It means that the algorithm classified Covid exams as normal ones. On the other hand, the achieved specificity was 80%, meaning that there are a considerable quantity of normal exams correctly classified.

3.2 *k-Nearest Neighbors (kNN)*

K-Nearest Neighbors (kNN) [7] is a machine learning algorithm used for classification and regression tasks. The main idea is to find the K nearest neighbors in the feature space to a given data point and use those neighbors to make a prediction.

Table 1 Nearest Centroid (NC) machine learning technique—performance metrics (mean and standard deviation)

Performance metrics	Nearest Centroid (NC)
Accuracy (Mean)	0.599
Accuracy (Std)	$1.11e^{-16}$
Specificity (Mean)	0.86
Specificity (Std)	0.00
Sensitivity (Mean)	0.339
Sensitivity (Std)	$5.55e^{-17}$
Gmean (Mean)	0.540
Gmean (Std)	0.00
AUC (Mean)	0.60
AUC (Std)	$1.11e^{-16}$

Table 2 k-Nearest Neighbors (kNN) machine learning technique—performance metrics (mean and standard deviation)

Performance metrics	k-Nearest Neighbors (kNN)
Accuracy (Mean)	0.805
Accuracy (Std)	0.00
Specificity (Mean)	0.960
Specificity (Std)	$2.22e^{e-16}$
Sensitivity (Mean)	0.650
Sensitivity (Std)	$1.11e^{e-16}$
Gmean (Mean)	0.789
Gmean (Std)	$1.11e^{e-16}$
AUC (Mean)	0.804
AUC (Std)	0.00

In kNN, for each new data point, the algorithm calculates the distance between that point and all the points in the training dataset. The K closest data points are then used to predict the target variable, either by taking the majority class in the case of classification or by taking the average value in the case of regression. Moreover, kNN does not require a training phase, which makes it fast and efficient.

Table 2 illustrates the Accuracy, Specificity, Sensibility, Gmean, and AUC with their respective mean and standard deviation (Std) for the k-Nearest Neighbors (kNN) classifier.

The KNN classifier presented regular performance metrics results, achieving accuracy, Specificity, and AUC of 80.5%, 96%, and 80.4%, respectively. On the other hand, it achieves a sensitivity of 65%, meaning that there was a considerable quantity of Covid exams classified as normal.

3.3 *Support Vector Machines (SVM)*

Another distance-based machine learning algorithm is the Support Vector Machines [6]. It is based on the concept of finding a hyperplane that separates the data into different classes in the case of classification or predicts the target variable in the case of regression. The algorithm seeks to find the hyperplane with the largest margin between the classes, which is the distance between the closest data points of different classes. The points closest to the hyperplane are called support vectors, and they influence the hyperplane's position.

Table 3 illustrates the Accuracy, Specificity, Sensibility, Gmean, and AUC with their respective mean and standard deviation (Std) for the Support Vector Machines (SVM) classifier.

3.4 *Random Forest (RF)*

Random Forest classifier [2] is an ensemble methodology. An ensemble method is a technique where multiple models are combined to make a single prediction. In the case of Random Forest, multiple decision trees are combined to make a final prediction, either by taking the average of the predictions or by taking a majority vote. The algorithm works by randomly selecting data points and features to split the data and build multiple decision trees. This makes the model robust to overfitting and improves its accuracy compared to a single decision tree.

Table 4 illustrates the Accuracy, Specificity, Sensibility, Gmean, and AUC with their respective mean and standard deviation (Std) for the Random Forest (RF) classifier.

Table 3 Support Vector Machines (SVM) machine learning technique—performance metrics (mean and standard deviation)

Performance metrics	Support Vector Machines (SVM)
Accuracy (Mean)	0.90
Accuracy (Std)	0
Specificity (Mean)	0.975
Specificity (Std)	0
Sensitivity (Mean)	0.825
Sensitivity (Std)	0
Gmean (Mean)	0.896
Gmean (Std)	0
AUC (Mean)	0.9
AUC (Std)	$1.1e^{-16}$

Table 4 Random Forest (RF) machine learning technique—performance metrics (mean and standard deviation)

Performance metrics	Random Forest (RF)
Accuracy (Mean)	0.768
Accuracy (Std)	0.005
Specificity (Mean)	0.938
Specificity (Std)	0.005
Sensitivity (Mean)	0.599
Sensitivity (Std)	0.010
Gmean (Mean)	0.749
Gmean (Std)	0.006
AUC (Mean)	0.768
AUC (Std)	0.005

The RF classifier presented regular performance metrics, achieving accuracy, Specificity, and AUC of 76.8%, 93.8%, and 76.4%, respectively. On the other hand, it achieves a sensitivity of 59.9%, meaning that a considerable quantity of Covid exams was classified as normal.

3.5 *Histogram Gradient Boosting (HGB)*

The Histogram Gradient Boosting classifier [5] is based on the gradient boosting framework, which builds a series of simple models called decision trees, and combines them to form a single, more complex model.

In Histogram Gradient Boosting, instead of using a traditional decision tree, which splits the data into smaller subsets based on a threshold for a single feature, the algorithm uses a histogram-based approach. The histogram-based approach splits the data into multiple bins based on the values of the feature and then calculates the gradient of the target variable with respect to the feature in each bin. This gradient information is then used to determine the best split for each tree in the series.

Table 5 illustrates the Accuracy, Specificity, Sensibility, Gmean, and AUC with their respective mean and standard deviation (Std) for the Histogram Gradient Boosting (HGB) classifier.

The HGB classifier presented good performance metrics, achieving accuracy, Specificity, and AUC of 84.8%, 96.3%, and 84.8%, respectively. On the other hand, it achieves a sensitivity of 73.4%, meaning that some Covid exams were classified as normal.

Table 5 Histogram Gradient Boosting (HGB) machine learning technique—performance metrics (mean and standard deviation)

Performance metrics	Histogram Gradient Boosting (HGB)
Accuracy (Mean)	0.848
Accuracy (Std)	0.008
Specificity (Mean)	0.963
Specificity (Std)	0.005
Sensitivity (Mean)	0.734
Sensitivity (Std)	0.017
Gmean (Mean)	0.840
Gmean (Std)	0.009
AUC (Mean)	0.848
AUC (Std)	0.008

3.6 Consolidated Results

In Table 6 are presented the performance metrics Accuracy, Specificity, Sensibility, Gmean, and AUC with their respective mean and standard deviation for each evaluated classifier.

SVM classifier outperforms the other algorithms in this experimental design, obtaining metrics of 90% for accuracy, 97.5% of Specificity, 82.5% of Sensitivity, 89.6% of Geometric mean, and 90% for the AUC. GTB and kNN presented regular performance, achieving acceptable results in terms of our evaluated metrics. Note that the Nearest Centroid and Random forest classifiers presented lower results in terms of sensitivity, meaning that there are Covid exams miss-classified in the test phase.

Table 6 Machine learning feature-based classifiers performance evaluation

Algorithm	Accuracy mean (std)	Specificity mean (std)	Sensitivity mean (std)	Gmean mean (std)	AUC mean (std)
NC	0.59 ($1.1e^{-16}$)	0.86 (0)	0.339 ($5.5e^{-17}$)	0.540 (0)	0.6 ($1.1e^{-16}$)
KNN	0.805 (0)	0.96 ($2.2e^{-16}$)	0.65 ($1.1e^{-16}$)	0.789 ($1.1e^{-16}$)	0.804 (0)
SVM	**0.90 (0)**	**0.975 (0)**	**0.825 (0)**	**0.896 (0)**	**0.9** ($1.1e^{-16}$)
RF	0.768 (0.005)	0.938 (0.005)	0.599 (0.010)	0.749 (0.006)	0.768 (0.005)
GTB	0.848 (0.008)	0.963 (0.005)	0.734 (0.017)	0.840 (0.009)	0.848 (0.008)

4 Conclusions

In this chapter, we proposed a pipeline for feature extraction and classification of Covid-19 in X-ray images using machine learning classifiers. In the proposed pipeline, five classifiers were evaluated through Accuracy, Specificity, Sensitivity, Geometric mean, and Area under the curve. The SVM Classifier presented the best performance metrics for Covid-19 classification, achieving 90% Accuracy, 97.5% of Specificity, 82.5% of Sensitivity, 89.6% of Geometric mean, and 90% for the AUC metric. On the other hand, the Nearest Centroid (NC) classifier presented poor sensitivity and geometric mean results, achieving 33.9% and 54.07%, respectively.

Our proposed pipeline uses transfer learning for feature extraction with VGG-16 CNN and classification with five classifiers. The Holdout and inner five-fold stratified cross-validation with grid search demonstrated competitive results.

Furthermore, this chapter did not fully explore the potential of COVID-19 classification using feature-based classifiers. In future work, we can experiment with other feature extraction techniques, CNNs, and hand-engineered features. It is also advisable to perform a feature selection analysis to handle the high-dimensional data generated by the extractors. Additionally, we can evaluate more classifiers aiming to achieve better results from our methodology.

Acknowledgements Bruno Riccelli dos Santos Silva would like to thank the support of the Coordination for the Improvement of Higher Education Personnel—Brazil (CAPES)—Financing Code 001.
Paulo Cesar Cortez would like to thank the Brazilian National Council for Scientific and Technological Development (CNPq) under Grant No. 313599/2019-0.
Joao Alexandre Lobo Marques would like to thank the research team from the Laboratory of Applied Neurosciences at the University of Saint Joseph, Macau SAR, China.

References

1. Aboutalebi H, Pavlova M, Gunraj H, Shafiee MJ, Sabri A, Alaref A, Wong A (2022) Medusa: multi-scale encoder-decoder self-attention deep neural network architecture for medical image analysis. Front Med 8:2891
2. Breiman L (2001) Random forests. Mach Learn 45:5–32
3. Breve FA (2022) Covid-19 detection on chest x-ray images: a comparison of cnn architectures and ensembles. Expert Syst Appl 204:117549
4. Cao Z, Huang J, He X, Zong Z (2022) Bnd-vgg-19: a deep learning algorithm for covid-19 identification utilizing x-ray images. Knowl-Based Syst 258:110040
5. Chen T, Guestrin C (2016) Xgboost: a scalable tree boosting system. In: Proceedings of the 22nd acm sigkdd international conference on knowledge discovery and data mining, pp 785–794
6. Cortes C, Vapnik V (1995) Support-vector networks. Mach Learn 20:273–297
7. Cover T, Hart P (1967) Nearest neighbor pattern classification. IEEE Trans Inf Theory 13(1):21–27
8. da Silva Neto MG, do Vale Madeiro JP, Gomes DG (2022) On designing a biosignal-based fetal state assessment system: a systematic mapping study. Comput Methods Progr Biomed 216:106671

9. da Silva Neto MG, do Vale Madeiro JP, Marques JAL, Gomes DG (2021) Towards an efficient prognostic model for fetal state assessment. Measurement 185:110034
10. Hussain L, Nguyen T, Li H, Abbasi AA, Lone KJ, Zhao Z, Zaib M, Chen A, Duong TQ (2020) Machine-learning classification of texture features of portable chest x-ray accurately classifies covid-19 lung infection. BioMed Eng OnLine 19:1–18
11. Hussain MdG, Shiren Y (2021) Recognition of covid-19 disease utilizing x-ray imaging of the chest using cnn. In: 2021 international conference on computing, electronics & communications engineering (iCCECE). IEEE, pp 71–76
12. Marques JAL, Fong SJ (eds) (2022) Epidemic analytics for decision supports in COVID-19 crisis. Springer International Publishing, Cham
13. Marques JAL, Gois FNB, do Vale Madeiro JP, Li T, Fong SJ (2022) Chapter 4 - Artificial neural network-based approaches for computer-aided disease diagnosis and treatment. In: Bhoi AK, de Albuquerque VHC, Srinivasu PN, Marques G (eds), Cognitive and soft computing techniques for the analysis of healthcare data. Intelligent data-centric systems. Academic, pp 79–99
14. Marques JAL, Gois FNB, da Silveira JAN, Li T, Fong SJ (2022) Chapter 5 - AI and deep learning for processing the huge amount of patient-centric data that assist in clinical decisions. In: Bhoi AK, de Albuquerque VHC, Srinivasu PN, Marques G (eds), Cognitive and soft computing techniques for the analysis of healthcare data. Intelligent data-centric systems. Academic, pp 101–121
15. Ohata EF, Bezerra GM, das Chagas JVS, Neto AVL, Albuquerque AB, de Albuquerque VHC, Filho PPR (2021) Automatic detection of covid-19 infection using chest x-ray images through transfer learning. IEEE/CAA J Autom Sinica 8(1):239–248
16. Pavlova M, Terhljan N, Chung AG, Zhao A, Surana S, Aboutalebi H, Gunraj H, Sabri A, Alaref A, Wong A (2022) Covid-net cxr-2: an enhanced deep convolutional neural network design for detection of covid-19 cases from chest x-ray images. Front Med 9
17. Tibshirani R, Hastie T, Narasimhan B, Chu G (2002) Diagnosis of multiple cancer types by shrunken centroids of gene expression. Proc Natl Acad Sci 99(10):6567–6572
18. Yousefpanah K, Ebadi MJ (2022) Review of artificial intelligence-assisted covid-19 detection solutions using radiological images. J Electron Imaging 32(2):021405
19. Zhao W, Jiang W, Qiu X (2021) Fine-tuning convolutional neural networks for covid-19 detection from chest x-ray images. Diagnostics 11(10):1887

Classification of COVID-19 CT Scans Using Convolutional Neural Networks and Transformers

Francisco Nauber Bernardo Gois, Joao Alexandre Lobo Marques, and Simon James Fong

Abstract COVID-19 is a respiratory disorder caused by CoronaVirus and SARS (SARS-CoV2). WHO declared COVID-19 a global pandemic in March 2020 and several nations' healthcare systems were on the verge of collapsing. With that, became crucial to screen COVID-19-positive patients to maximize limited resources. NAATs and antigen tests are utilized to diagnose COVID-19 infections. NAATs reliably detect SARS-CoV-2 and seldom produce false-negative results. Because of its specificity and sensitivity, RT-PCR can be considered the gold standard for COVID-19 diagnosis. This test's complex gear is pricey and time-consuming, using skilled specialists to collect throat or nasal mucus samples. These tests require laboratory facilities and a machine for detection and analysis. Deep learning networks have been used for feature extraction and classification of Chest CT-Scan images and as an innovative detection approach in clinical practice. Because of COVID-19 CT scans' medical characteristics, the lesions are widely spread and display a range of local aspects. Using deep learning to diagnose directly is difficult. In COVID-19, a Transformer and Convolutional Neural Network module are presented to extract local and global information from CT images. This chapter explains transfer learning, considering VGG-16 network, in CT examinations and compares convolutional networks with Vision Transformers (ViT). Vit usage increased VGG-16 network F1-score to 0.94.

F. N. Bernardo Gois · J. A. Lobo Marques (✉)
Laboratory of Applied Neurosciences, University of Saint Joseph,
Estrada Marginal da Ilha Verde, 14-17, Macao SAR, China
e-mail: alexandre.lobo@usj.edu.mo

F. N. Bernardo Gois
e-mail: nauber.gois@usj.edu.mo

S. J. Fong
Faculty of Science and Technology, University of Macau, Macau SAR, China
e-mail: ccfong@um.edu.mo

J. A. Lobo Marques and S. J. Fong (eds.), *Computerized Systems for Diagnosis and Treatment of COVID-19*, https://doi.org/10.1007/978-3-031-30788-1_6

1 Introduction

COVID-19 (Corona Virus Disease 2019) is a respiratory disease caused by SARS-CoV2 virus.The World Health Organization (WHO) proclaimed COVID-19 a pandemic in March 2020. This has brought the healthcare system of different nations to the operational edge, with some specific regions facing collapses. Therefore, precisely screening COVID-19-positive patients to use scarce resources efficiently is of the utmost importance. Popularly used to diagnose COVID-19 infections are the Nucleic Acid Amplification Tests (NAATs) and Antigen Tests. NAATs can detect SARS-CoV-2 consistently and are unlikely to produce a false-negative result. The most popular test for COVID-19 is Reverse Transcription Polymerase Chain Reaction (RT-PCR) because of its high levels of specificity (true negative rate) and sensitivity (true positive rate). However, this test is expensive and time-consuming due to its sophisticated kit. Using nasal or throat swabs to detect SARS-CoV-2, an RT-PCR test is conducted by skilled specialists who have been instructed on how to use the RT-PCR kit. For achieving the result, RT-PCR requires a full setup consisting of experienced practitioners, a laboratory, and an RT-PCR machine [46].

Since COVID-19 is a disease that commonly affects the patients' lungs with pneumonia in several different ways, medical imaging exams, such as X-Rays and CT-Scans of the chest, are commonly required by medical professionals to assess conditions and risks of patients.

Computerized systems based on Artificial Intelligence (AI) techniques are considered to support diagnosis for many different types of medical imaging applications. More recently, models known as Deep Learning Networks have been proposed to successfully extract information from images of Computerized Tomography (CT) of the chest as a new and efficient image processing technique. However, specifically for patients infected with COVID-19, the detectable lesions in the lungs images are widely dispersed and present various local aspects [16]. Therefore, using the current deep learning model to process a diagnosis of COVID-19 using CT-Scans of the chest becomes a challenging task.

To address this problem, a new system is proposed in this chapter, integrating a Convolutional Neural Network (CNN) from a Transfer Learning model and a Transformer Network, following the medical features of CT-Scan images of patients with COVID-19. This approach aims to use the best capability of CNNs for local feature extraction and the potential of Transformer Networks for global feature extraction.

To evaluate the performance of the proposed system, Transfer Learning is considered to use a model for classifying CT Scan exams. The VGG-16 model is evaluated, achieving an F1-score of 0.91, while the proposed system presented a better performance with an F1-score of 0.94.

2 Concepts and Technical Background

Deep Learning models have been exponentially used to detect respiratory diseases in the literature [52], and, motivated by the impacts of the COVID-19 pandemic, several works are using different Deep Learning models to assess CXR (Chest X-Ray) and CCT (Chest Computerized Tomography) as an efficient approach to diagnose COVID-19 cases [34, 73]. In contrast to a healthy individual's CCT/CXR, the COVID-19 patients' images present remarkable changes, such as ground-glass opacity (GGO) and consolidations all over the lungs [66, 73].

While there is a considerable body of research on the application of Deep Learning for Covid identification, a significant number of studies are based on Convolutional Neural Networks (CNNs) [23, 67]. Due to its image-specific inductive biases, CNN, despite its strength, lacks a universal comprehension of pictures. To capture long-range relationships, CNNs require a huge receptive field, which involves constructing enormous kernels or incredibly deep networks, resulting in a complicated model that is tough to train. A new framework called Vision transformers [15] has been recently presented for learning tasks with the aim to overcome issues associated with CNN inductive bias for specific classification tasks.

2.1 Transformers

Transformers Networks [65] are Deep Learning models that have already been adopted in multiple domains, including speech recognition or natural language, and medical image processing. Transformers Networks were initially presented as a sequential model that could be used in translation tasks [31]. Following that development, Qiu et al. demonstrate that Transformer-based pre-trained models can be applied to multiple problems, achieving and surpassing the performance available in the scientific literature [50]. As a result, these networks became the standard model for natural language problems. In addition, Transformers networks are also very efficient for Computer Vision applications [8, 15, 49], audio processing [10, 14, 22], and even other fields, such as chemistry [57] and life sciences [42, 55].

In recent years, numerous Transformer variants (also known as X-formers) have been proposed due to the popularity achieved by the technique. These X-formers enhance the standard Transformer in various ways, and the Vanilla Transformer is presented in the following Subsection.

2.2 Vanilla Transformer

The vanilla Transformer [65] utilizes an encoder-decoder approach, to create a sequential model with multiple blocks. Encoder blocks are designed to create a multi-head self-attention module and a position-wise feed-forward network (FFN).

A residual connection [25] is utilized around each module, followed by a Layer Normalization [4] module for developing a deeper model. In addition to inserting cross-attention modules between the multi-head self-attention modules and the position-wise FFNs, decoder blocks also include cross-attention modules. In addition, the decoder's self-attention modules are modified to prevent each position from attending to succeeding positions.

2.3 *Transformer Architecture*

The foundational ideas for building Transformer architectures are on a mechanism of self-attention, which aims to discover links between a sequence of elements. In contrast to recurrent networks, which process sequence elements recursively and can only attend to short-term context, Transformers can attend to whole sequences, thereby learning long-range correlations. The Transformers approach is based on a unique implementation, such as the structure called multi-head attention, optimized for execution in parallel. They are designed to answer to problems with higher complexity, necessity for scalability, and large datasets, including Big Data problems.

The Transformers networks have a pre-training phase using unlabeled large-scale datasets [13, 65] since one basic assumption of these systems is the low level of prior knowledge of the problem, if compared with recurrent ANNs or convolutional Deep Learning models [20, 21, 38]. This pre-training phase encodes data representation and relationships from the dataset under consideration, avoiding the need for costly manual annotations, which are actually not available for real-life huge databases. The representations are considered by the system and the model goes to the fine-tuning phase with subsequent supervised learning techniques to achieve improved results [9].

In summary, the successful development of Transformer-based models relies on two main characteristics: Firstly, the self-attention module, which enables the capture of "long-term" dependencies and relationships between data sequence elements, which is not considered when using recurrent models. Secondly, the self-supervised pre-training phase on using a large unlabeled dataset, followed by a fine-tuning phase focused on the target task with small labeled sub-datasets [13, 45, 68]. The two concepts are explained in more detail in the following paragraphs, followed by a summary of Transformer networks in which these concepts have been applied. This context will aid our comprehension of forthcoming computer vision Transformer models.

2.3.1 Self-attention Mechanism

The self-attention mechanism consists on employing a two-stage training mechanism. First, supervised [15] or unsupervised [13, 40, 41] pre-training is performed on a large-scale dataset (and sometimes a combination of several available datasets

[11, 59]. The pre-trained weights are subsequently adapted to downstream tasks using small-to-medium scale datasets. Image classification [18], object detection [8], zero-shot classification [54], question answering [17], and action recognition [19] are examples of downstream tasks. Pre-training for large-scale Transformers has been deemed effective in both the language and vision domains. For instance, the Vision Transformer model (ViT-L) [15] experiences a 13% decrease in accuracy on the ImageNet test set when trained solely on the ImageNet train set as opposed to when pre-trained considering the 300 million-image JFT dataset [60].

2.3.2 Pre-training and Fine-Tuning

The task of labeling large amounts of data is time-consuming and a source of manual identification errors. Transformer networks consider self-supervised learning for a pre-training phase, making it a scalable model. For example, the Switch Transformer from Google considers the use of more than one trillion parameters [17]. As another example is the Bibliography, [33, 43], aims to predict occluded data in images, future or past frames in temporal video sequences. Finally, image processing using these models is presented in [26].

Another application is known as contrastive learning and presents significant benefits of using self-supervised models. This learning technique generates two different images from the original one. The first image (transformed image 1) keeps the original class categorization, while the second image (transformed image 2) removes the class categorization. The model is trained to be invariant to minor changes, such as the ones from transformed image 1, creating a focus on modifications that can alter the class categorization, which is present in transformed image 2.

2.3.3 Additional Characteristics

Some relevant challenges faced by Transformers networks are presented as follows:

- Model Efficiency. The self-attention module deals with large amounts of data and high demand for processing power, so processing and memory capacities are required. Because of that, the inefficiency of a Transformer when processing lengthy sequences is a significant obstacle to its application. Lightweight attention (such as sparse attention versions) and Divide-and-conquer techniques are among the enhancement methods.
- Model Generalization. Small-scale data is challenging to train because the transformer has a flexible design and makes few assumptions about the structural bias of incoming data. The strategies for improvement include incorporating structural bias or regularization, the pre-training phase using massive unlabeled data, etc.
- Model Adaptation. The challenge to adapt existing Transformer networks to achieve the resolution of specific tasks is a complex task.

2.4 Covid-19 Detection Using CT Scans of the Chest

Alternative COVID-19 screening methods include chest CT imaging [5, 73]. Several attributes, such as Volume, Radiomics characteristics, Infected lesion number, Histogram distribution, and Surface area, are extracted from CT images in [61], followed by discriminative feature selection and classification using a deep forest algorithm with cascaded layers of multiple random forests.

It is essential to highlight that extracting radiomics characteristics can be considered a significantly valuable tool for obtaining relevant information from medical images. Some of the possible obtained attributes are First Order Statistics, 2D and 3D Shape-based features, and gray-level metrics. They can be used to classify image attributes, such as lesions or tumors, and, following the analysis of the patient's other symptoms and general conditions, predict possible outcomes.

A comprehensive work in [3] optimizes 10 pre-trained CNN Transfer Learning models, including AlexNet [36], VGG-16 [58], VGG-19 [58], SqueezeNet [32], GoogleNet [62], MobileNet-V2 [56], ResNet-18 [25], ResNet-50 [25], ResNet-101 [25], and Xception [12], on CT-scan images to differentiate COVID-19 instances from non-cases.

According to [3], ResNet-101 and Xception perform best. COVNet [39], a CNN architecture based on ResNet50, receives a sequence of CT slices as input and computes features from each slice, which are merged via a max-pooling operation and supplied to a fully connected layer to provide a probability score for each class. A multilayer perceptron (MLP) aggregates slice predictions to generate a patient prediction. However, COVIDNet-CT [23] offers architectural flexibility, customizable long-range connection, and lightweight design patterns. Contrastive COVIDNet [70] is based on COVIDNet [67] and adds domain-specific batch normalization layers, cross-entropy classification, and contrastive loss. In [48], a CNN model with two forward pass lines and deep feature aggregation is created to distinguish COVID from non-COVID. CT and X-ray data can be processed by the network. It uses deep feature aggregation to combine layer outputs from different depths after a classifier network. ResGNet-C [74] employs Graph Convolution Network (GCN) [35] to perform binary classification tasks with Resnet-101 [25] generated features.

2.5 Covid-19 Detection Using Transformers

Google introduced Transformer Deep Learning Models [65] in 2017 for natural language processing. The Transformer network structure is mostly attention processes with a global receptive field. "Transformers" is a global CNN. Encoders and decoders make up transformers. Transformer's Self-Attention module use normalized dot product attention. Dot multiplication on all q and k is the Self-Attention mechanism. Before that, normalize every input mapping to get the attention weight matrix. Assuming that the input dimensions of query and key are d_k, and the value dimension is d_v (since the Transformer structure was first applied in natural language processing, symbols like query, key, and value continue to be used). The point-multiplied operation of query and each key is calculated and divided by $\sqrt{d_k}$, and then the weight is calculated by softmax function [16].

$$Attention(Q, K_i, V_i) = softmax\left(\frac{Q^t K_i}{\sqrt{d_k}}\right) V_i \tag{1}$$

The encoder has two layers of Norm and Add, which are represented by the equations:

$$\begin{gathered} LayerNorm(X + MultiHeadAttention(X)) \\ LayerNorm(X + FeedForward(X)) \end{gathered} \tag{2}$$

The dimensions of the output (X) and input (X) of Multi-Head Attention or Feed Forward are the same, thus, they can be added. X is the input of Multi-Head Attention or Feed Forward. Two layers that are fully connected connect to the Feed Forward layer. Relu serves as the first layer's activation function, while the activation function of the second layer is not utilized. These are the equivalent formulas [16].

Using CT and X-ray images, [2] proposes a novel deep-learning framework for coronavirus detection. The proposed network specifically adopts a Vision Transformer architecture as its backbone and uses a Siamese encoder. It has two branches: one for processing the original image and one for processing an enhanced version of the original image. After being separated into patches, the input images are sent to the encoder. Considering open CT and X-ray datasets, the suggested framework is assessed. On CT and X-ray data, the suggested approach demonstrates its advantages over cutting-edge techniques in terms of accuracy, precision, recall, specificity, and F1 score. The suggested approach also shows promising robustness when only a modest amount of training data is distributed.

In order to solve this problem, [27] concurrently proposes 2-D and 3-D models to forecast the COVID-19 of CT scans. To address the previously noted problem, the study incorporates the Deep Wilcoxon signed-rank test (DWCC) into our 2-D model to assess the significance of each CT scan slice. A Convolutional CT scan-Aware Transformer (CCAT) is also suggested to properly understand the context of the slices. Each CT slice is considered for extracting the frame-level feature based on any backbone network. The features are then sent to our within-slice-Transformer (WST), which finds the context information in the pixel dimension. The retrieved spatial-context features of each CT slice are aggregated using the suggested BetweenSlice-Transformer (BST). The COVID-19 status of the spatiotemporal features is then determined by a straightforward classifier.

2.6 Visual Transformers

Transformers [65], as previously presented, are networks using self-attention which are being widely used for natural language and sequence modeling tasks. Some recent approaches are considering a self-supervised pre-training phase with large amounts of unlabeled data followed by a fine-tuning phase with labeled data for particular tasks [47, 51].

Another approach considers the family of Generative Pre-trained Transformer (GPT) [6, 51], which focus on using decoders to enable natural language processing. Other families of models are the BERT—Bidirectional Encoder Representations from Transformers from which different applications can be found in the literature [13, 37, 44, 76].

Before the visual Transformers, Convolution Neural Networks (CNNs) have become the dominating paradigm in the computer vision area [36, 63]. The concept of self-attention has been proposed for other CNN-based models to address the same capability of detecting long-range relationships in large amounts of data at either spatial level [7, 30, 69] or channel level [29, 71]. In contrast, others attempted to change the convolutional approach with global [49] or local self-attention blocks [28, 64]. In another work, Ramachandr et al. determined the effectiveness of self-attention blocks without using CNNs to assist the classification process [53]. Nevertheless, the performance of these sole attention models was lower than the CNN models previously presented [44].

By (a) encoding images as arrays of evenly ordered pixels and (b) convolving highly-localized characteristics, computer vision has achieved amazing success. However, convolutions treat all image pixels equally regardless of their significance, explicitly model all concepts across all images regardless of their content, and struggle to relate concepts that are spatially distant. In this study, we challenge this paradigm by (a) encoding images as semantic visual tokens and (b) executing transformers to model token connections in great detail. Visual Transformer (ViT) functions in a semantic token space, giving careful consideration to various visual components based on context. This is in stark contrast to pixel-space converters, which require power in a higher level of magnitude. Using an enhanced training recipe, ViTs outperform their convolutional counterparts by a large margin, increasing ResNet's accuracy on ImageNet top-1 by 4.6 to 7 points while increasing ResNet's accuracy by a comparable amount [72]. ViT has achieved state-of-the-art performance on various image recognition benchmarks [15]. In addition to basic image classification, the transformer has been applied to a number of additional computer vision issues, such as object recognition [8, 75], semantic segmentation, image processing, and video comprehension. As a result of its excellent performance, an increasing number of academics are proposing transformer-based models for enhancing a vast array of visual tasks [24].

2.7 Transfer Learning

Transfer learning is a machine learning method where a model developed for a task is reused as the starting point for a model on a second task. It is an increasingly popular strategy for applying deep learning approaches considering pre-trained models (architecture and weights) to be adopted as the starting point for achieving one task. Currently, the applications are mainly in the areas of image processing/computer vision and natural language processing. These applications demand heavy computing power (often with expensive GPU farms) for a long time, creating a burden of resources (computational and financial).

Three popular groups of Transfer Learning models available are from Oxford, known as VGG Models; from Google, called Inception Models; and from Microsoft, the ResNet Models. In this chapter, the VGG-16 model is considered, and the application is explained further.

2.8 Evaluation Criteria

This chapter will consider the confusion matrix and three performance indicators to evaluate the proposed classification systems: Accuracy, Precision, and Recall.

2.8.1 Confusion Matrix

First, the Confusion Matrix is a common representation for the evaluation of false negatives and false positives identified by the proposed classifier. More formally, the matrix can be stated as follows.

Let MI(x,y) : $\mathbb{R}^2 \rightarrow \mathbb{R}$ represent a medical image and O(MI(x,y)):$\mathbb{R}^2 \rightarrow \Omega, \Omega =$ 0, 1 a classification of the medical image MI(x,y). The classification criteria is then presented considering the classification goal as G, and the result achieved by the system as R [?]:

- true positive: $G(x, y) = 1$ ^$R(x, y) = 1$,
- false positive: $G(x, y) = 0$ ^$R(x, y) = 1$,
- true negative: $G(x, y) = 0$ ^$R(x, y) = 0$,
- false negative: $G(x, y) = 1$ ^$R(x, y) = 0$.

2.8.2 Accuracy

Accuracy is a popular metric for evaluating a model's performance, considering both positive and negative classes and true and false positives (TP and FP), and true and false negatives (TN and FN) in the same metric. The equation is presented as follows:

$$Accuracy = \frac{TP + TN}{TP + TN + FP + FN}$$

2.8.3 Precision

The Precision is a metric to evaluate the system performance from the perception of the number of false positives, compared to the number of true positive cases and is given by:

$$P = \frac{TP}{TP + FP}$$

2.8.4 Recall

The Recall (R) is a metric to evaluate the system performance from the perception of the number of false negatives (FN), compared to the number of true positive (TP) cases, i.e., it measures the system performance not to miss one TP case and is given by:

$$R = \frac{TP}{TP + FN}$$

3 Experimental Results and Discussion

In this chapter, we suggest a test comparing the use of transformers and convolutional networks for classifying CT-Scan images. The in-depth description of the database provided by Union Hospital (HUST-UH) and Liyuan Hospital (HUST-LH) [1]. In this study, the total number of 19,685 images of individual CT scans was divided into three categories:

- 5,705 non-informative CT (NiCT) images, in which lung parenchyma was not captured for any judgment;
- 4,001 positive CT (pCT) images (Fig. 1), in which imaging features associated with COVID-19 pneumonia could be clearly identified;
- 9,979 negative CT (nCT) images, in which imaging features in both lungs were unrelated to COVID-19 pneumonia

Therefore, the initial methodology described in the research was removing lung parenchyma from CT images. Using the techniques described in the publication, we retrieved lung parenchyma. The original images are in the original CT scans folder, while the extracted images are in the preprocessed CT scans folder. Every image has been downsized to (512×512).

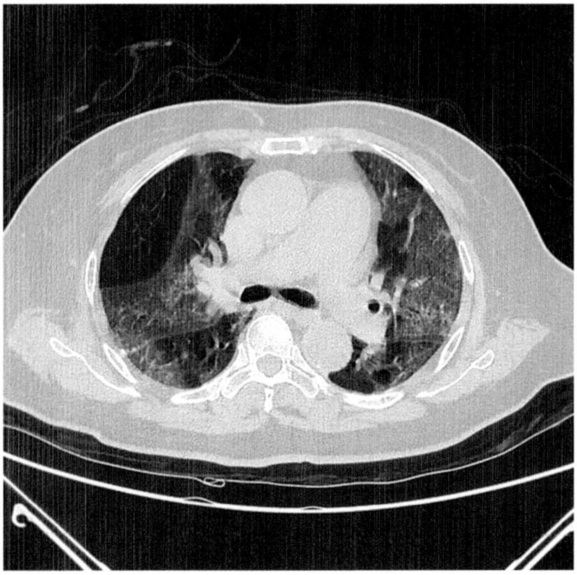

Fig. 1 Figure with a CT-Scan image of a patient with a positive diagnosis for COVID [1]

The experiment of this chapter aims to conduct a comparative study between the use of VGG-16 with transfer learning and a visual transformer network (ViT). ViT, applies a Transformer-like design to selected areas of the image. A sequence of vectors is created by dividing a picture into fixed-size patches, linearly embedding each one, adding position embeddings, and then feeding the assembled vectors to a conventional Transformer encoder. The traditional classification method involves including an extra learnable "classification token" in the sequence.

Compared to CNNs, Vision Transformer exhibits substantially less image-specific inductive bias. In CNN's, each layer of the entire model is baked with locality, two-dimensional neighborhood structure, and translation equivariance. In ViT, the self-attention layers are global, but only the MLP levels are local and translationally equivariant. When fine-tuning the model, the position embeddings are changed to account for images of various resolutions, and only very infrequently is the two-dimensional neighborhood structure used (as described below). Other than that, all spatial interactions between the patches must be learned from the start because the position embeddings at initialization time include no information about the 2D positions of the patches.

Transformers measure what is known as attention-the connections between pairs of input tokens (words in the case of text strings). The cost is a quadratic function of the token count. The pixel serves as the fundamental analytical unit for images. However, the amount of memory and processing required to compute relationships for each pair of pixels in a typical image is prohibitive. Instead, at a far lower cost, ViT computes relationships between pixels in various small portions of the image (such as 16×16 pixels). The portions are organized into a sequence using positional

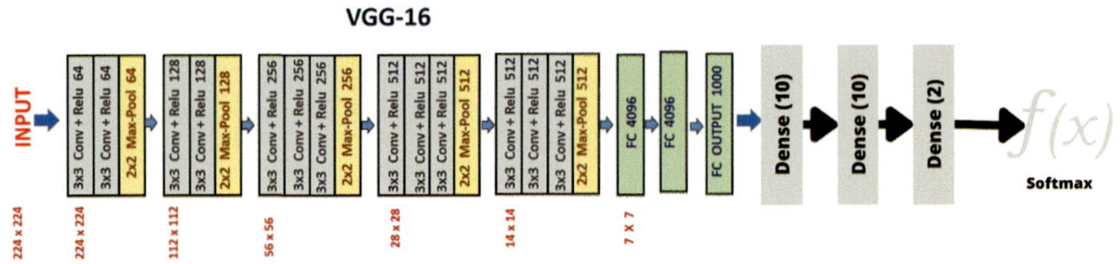

Fig. 2 VGG-16 network used in the experiment

embeddings. The vector embeddings can be learned. Each portion is multiplied by the embedding matrix and put in a linear order. The transformer receives the outcome together with the position embedding.

5250 images of 200 by 200 were selected for training and 1750 for testing. We use a VGG-16 Transfer Learning model with Imagenet weights (Fig. 2). In VGG-16, there are thirteen convolutional layers, five Max Pooling layers, and three Dense layers for a total of twenty-one layers, yet there are only sixteen weight layers, also known as the learnable parameters layer. The VGG-16 network input tensor size is 224, 244, possible to execute 3 RGB channels. The VGG-16 has convolution layers of 3×3 filters with stride 1 and always used the same padding and MaxPool layers of 2×2 filters with stride 2. This allows the reduction of hyperparameters. The convolution and maximum pool layers are ordered uniformly across the whole architecture.

The architecture of the VGG-16 networks considered in this work is presented as follows.

- Conv-1: 64 filters;
- Conv-2: 128 filters;
- Conv-3: 256 filters;
- Conv-4: 512 filters;
- Conv-5: 512 filters;
- Conv-6: 512 filters;
- Fully-Connected-1: 4096 channels
- Fully-Connected-2: 4096 channels
- Fully-Connected-3: 1000 channels
- Softmax.

For the Visual Transformer (ViT) Network, the parameters are:

- Patch size: 6×6
- Patches per image: 64
- Elements per patch: 108

Table 1 presents the performance metrics Precision, Recall, and F1-score for the VGG-16 Transfer Learning CNN.

In Fig. 3 is presented one example of a lung CT-Scan image, split into 64 patches for processing through the Transformer Network.

Table 1 Performance results obtained for the classification using the Transfer Learning CNN VGG-16

	Precision	Recall	f1-score	Support
0	0.88	0.94	0.91	893
1	0.93	0.87	0.90	857
Accuracy			0.91	1750
Macro avg	0.91	0.91	0.91	1750
Weighted avg	0.91	0.91	0.91	1750

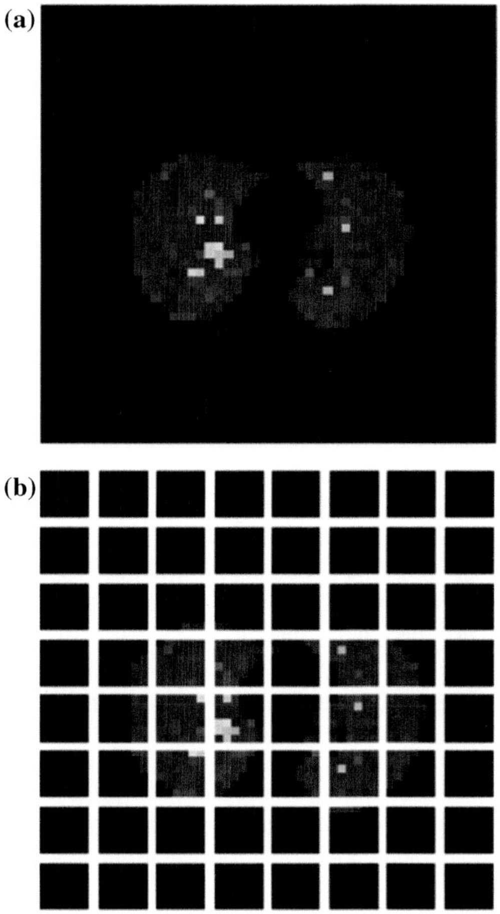

Fig. 3 Case representation of **a** CT-Scan image of the Chest with segmented lungs, and **b** with 64 patches for use in Visual Transformer

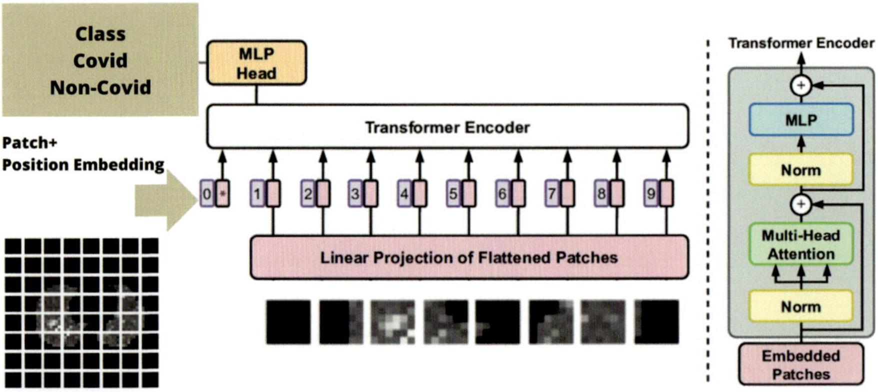

Fig. 4 Visual Transform (ViT) proposed for the classification of COVID-19 using CT Scan images experiment

ViT performance is contingent on a variety of decisions, including those related to the optimizer, dataset-specific hyperparameters, and network depth. CNN is significantly easier to optimize. The standard ViT stem employs a 16×16 convolution model with a 16 stride. In comparison, a 3×3 convolution with stride 2 enhances both stability and precision. The convolutional network turns simple pixel data into a map of features. A tokenizer converts the feature map into a sequence of tokens that are then passed into the transformer, which employs the attention mechanism to generate a sequence of output tokens. A projector is then used to rejoin the output tokens to the feature map. This permits the study to make use of potentially substantial pixel-level features. This dramatically minimizes the number of tokens that must be evaluated, hence decreasing costs. Figure 4 shows a comprehensive block diagram of the Visual Transformer Network proposed for the current experiment.

There are numerous variations between CNNs and Vision Transformers, most of which are architectural.

Comparing the two solutions, the literature indicates that CNN's may achieve superior performance when trained using significantly smaller datasets than those required by Vision Transformers. These results should be carefully analyzed since this outstanding performance may appear to result from the presence of certain inductive biases that can be utilized by these CNN's to grasp the particularities of the analyzed images more rapidly but at the expense of making it more challenging to comprehend global relationships or to generalize when submitted to larger and more comprehensive datasets. On the other hand, Vision Transformers are devoid of these biases, allowing them to catch global and broader range relations at the expense of more laborious training in data. Vision Transformers were also far more resistant to picture distortions such as hostile patches and permutations. Table 2 presents the performance metrics Precision, Recall and F1-score for the proposed Visual Transformer Network.

Table 2 Results obtained from ViT

	Precision	Recall	f1-score	Support
0	0.96	0.92	0.94	893
1	0.92	0.96	0.94	857
Accuracy			0.94	1750
Macro avg	0.94	0.94	0.94	1750
Weighted avg	0.94	0.94	0.94	1750

Considering the F1-Score metric as a reference and comparing Tables 1 and 2, it can be determined that the proposed Visual Transformer achieved 0.94, while the VGG-16 Transfer Learning CNN achieved 0.91 for the same metric.

4 Conclusion

In this chapter, different Artificial Intelligence models based on Deep learning networks have been proposed to extract medical information from chest CT images and as a novel clinical detection method to support the effective diagnosis of COVID-19. Analyzing CT scans of the chest and the extracted clinical features using conventional deep-learning methods is challenging because the images present spread lesions and have several local characteristics. To address this issue, a Transformer Convolutional Neural Network module is proposed to extract local and global information from CT scans and provide efficient classification. A comparison with a Transfer Learning CNN model (VGG-16) is provided While the CNN model achieved an F1-score of 0.91, the proposed Visual Transformer (ViT) achieved an F1-score of 0.94.

Future works might consider new approaches to compare with the proposed model. In addition, more data should be used to evaluate system performance. Finally, bringing the solution closer to the specialist analysis, implementing the model in a PACS (Picture Archiving and Communication System) to evaluate computational requirements and the feasibility of supporting medical diagnosis in an actual clinical setup.

References

1. Ct scans for covid-19 classification—kaggle. https://www.kaggle.com/datasets/azaemon/preprocessed-ct-scans-for-covid19. Accessed on 13 Nov 2022
2. Al Rahhal MM, Bazi Y, Jomaa RM, AlShibli A, Alajlan N, Mekhalfi ML, Melgani F (2022) Covid-19 detection in ct/x-ray imagery using vision transformers. J Personal Med 12(2):310

3. Ardakani AA, Kanafi AR, Acharya UR, Khadem N, Mohammadi A (2020) Application of deep learning technique to manage covid-19 in routine clinical practice using ct images: results of 10 convolutional neural networks. Comput Biol Med 121:103795
4. Ba JL, Kiros JR, Hinton GE (2016) Layer normalization. arXiv:1607.06450
5. Bernheim A, Mei X, Huang M, Yang Y, Fayad ZA, Zhang N, Diao K, Lin B, Zhu X, Li K et al (2020) Chest ct findings in coronavirus disease-19 (covid-19): relationship to duration of infection. Radiology
6. Brown T, Mann B, Ryder N, Subbiah M, Kaplan JD, Dhariwal P, Neelakantan A, Shyam P, Sastry G, Askell A et al (2020) Language models are few-shot learners. Adv Neural Inf Proc Syst 33:1877–1901
7. Cao Y, Xu J, Lin S, Wei F, Hu H (2019) Gcnet: non-local networks meet squeeze-excitation networks and beyond. In: Proceedings of the IEEE/CVF international conference on computer vision workshops, pp 0–0
8. Carion N, Massa F, Synnaeve G, Usunier N, Kirillov A, Zagoruyko S (2020) End-to-end object detection with transformers. In: European conference on computer vision. Springer, pp 213–229
9. Chaudhari S, Mithal V, Polatkan G, Ramanath R (2021) An attentive survey of attention models. ACM Trans Intell Syst Technol (TIST) 12(5):1–32
10. Chen X, Wu Y, Wang Z, Liu S, Li J (2021) Developing real-time streaming transformer transducer for speech recognition on large-scale dataset. In: ICASSP 2021-2021 IEEE international conference on acoustics, speech and signal processing (ICASSP). IEEE, pp 5904–5908
11. Chen Y-C, Li L, Yu L, El Kholy A, Ahmed F, Gan Z, Cheng Y, Liu J (2020) Uniter: universal image-text representation learning. In: European conference on computer vision. Springer, pp 104–120
12. Chollet F (2017) Xception: deep learning with depthwise separable convolutions. In: Proceedings of the IEEE conference on computer vision and pattern recognition, pp 1251–1258
13. Devlin J, Chang M-W, Lee K, Toutanova K (2018) Bert: pre-training of deep bidirectional transformers for language understanding. arXiv:1810.04805
14. Dong L, Xu S, Xu B (218) Speech-transformer: a no-recurrence sequence-to-sequence model for speech recognition. In: 2018 IEEE international conference on acoustics, speech and signal processing (ICASSP). IEEE, pp 5884–5888
15. Dosovitskiy A, Beyer L, Kolesnikov A, Weissenborn D, Zhai X, Unterthiner T, Dehghani M, Minderer M, Heigold G, Gelly S et al (2020) An image is worth 16x16 words: transformers for image recognition at scale. arXiv:2010.11929
16. Fan X, Feng X, Dong Y, Hou H (2022) Covid-19 ct image recognition algorithm based on transformer and cnn. Displays 102150
17. Fedus W, Zoph B, Shazeer N (2021) Switch transformers: scaling to trillion parameter models with simple and efficient sparsity
18. Gidaris S, Singh P, Komodakis N (2018) Unsupervised representation learning by predicting image rotations. arXiv:1803.07728
19. Girdhar R, Carreira J, Doersch C, Zisserman A (2019) Video action transformer network. In: Proceedings of the IEEE/CVF conference on computer vision and pattern recognition, pp 244–253
20. Goodfellow I, Bengio Y, Courville A (2016) Deep learning. MIT Press
21. Graves A (2012) Long short-term memory. In: Supervised sequence labelling with recurrent neural networks, pp 37–45
22. Gulati A, Qin J, Chiu C-C, Parmar N, Zhang Y, Yu J, Han W, Wang S, Zhang Z, Wu Y et al (2020) Conformer: convolution-augmented transformer for speech recognition. arXiv:2005.08100
23. Gunraj H, Wang L, Wong A (2020) Covidnet-ct: a tailored deep convolutional neural network design for detection of covid-19 cases from chest ct images. Front Med 7:608525
24. Han K, Wang Y, Chen H, Chen X, Guo J, Liu Z, Tang Y, Xiao A, Xu C, Xu Y et al (2020) A survey on visual transformer. 2(4). arXiv:2012.12556
25. He K, Zhang X, Ren S, Sun J (2016) Deep residual learning for image recognition. In: Proceedings of the IEEE conference on computer vision and pattern recognition, pp 770–778

26. Hinton G, LeCunn Y, Bengio Y (2020) Aaai'2020 keynotes turing award winners event
27. Hsu C-C, Chen G-L, Wu M-H (2021) Visual transformer with statistical test for covid-19 classification. arXiv:2107.05334
28. Hu H, Gu J, Zhang Z, Dai J, Wei Y (2018) Relation networks for object detection. In: Proceedings of the IEEE conference on computer vision and pattern recognition, pp 3588–3597
29. Hu J, Shen L, Sun G (2018) Squeeze-and-excitation networks. In: Proceedings of the IEEE conference on computer vision and pattern recognition, pp 7132–7141
30. Huang Z, Wang X, Huang L, Huang C, Wei Y, Liu W (2019) Ccnet: criss-cross attention for semantic segmentation. In: Proceedings of the IEEE/CVF international conference on computer vision, pp 603–612
31. Hya S, Oriol V, Quoc VL (2014) Sequence to sequence learning with neural networks. In: Conference on Advances in neural information processing systems
32. Iandola FN, Han S, Moskewicz MW, Ashraf K, Dally WJ, Keutzer K (2016) Squeezenet: alexnet-level accuracy with 50x fewer parameters and <0.5 mb model size. arXiv:1602.07360
33. Jing L, Tian Y (2020) Self-supervised visual feature learning with deep neural networks: a survey. IEEE Trans Pattern Anal Mach Intell 43(11):4037–4058
34. Khan AI, Shah JL, Bhat MM (2020) Coronet: a deep neural network for detection and diagnosis of covid-19 from chest x-ray images. Comput Methods Programs Biomed 196:105581
35. Kipf TN, Welling M (2016) Semi-supervised classification with graph convolutional networks. arXiv:1609.02907
36. Krizhevsky A, Sutskever I, Hinton GE (2017) Imagenet classification with deep convolutional neural networks. Commun ACM 60(6):84–90
37. Lan Z, Chen M, Goodman S, Gimpel K, Sharma P, Soricut R (2019) Albert: a lite bert for self-supervised learning of language representations. arXiv:1909.11942
38. LeCun Y, Bengio Y, Hinton G (2015) Deep learning. Nature 521(7553):436–444
39. Lin L, Lixin Q, Xu Z, Youbing Y, Xin W, Bin K, Junjie B, Lu Y, Zhenghan F, Qi S et al (2020) Using artificial intelligence to detect covid-19 and community-acquired pneumonia based on pulmonary ct: evaluation of the diagnostic accuracy. Radiology 296(2):E65–E71
40. Li X, Yin X, Li C, Zhang P, Hu X, Zhang L, Wang L, Hu H, Dong L, Wei F et al (2020) Oscar: object-semantics aligned pre-training for vision-language tasks. In: European conference on computer vision. Springer, pp 121–137
41. Lin K, Wang L, Liu Z (2021) End-to-end human pose and mesh reconstruction with transformers. In: Proceedings of the IEEE/CVF conference on computer vision and pattern recognition, pp 1954–1963
42. Lin T, Wang Y, Liu X, Qiu X (2022) A survey of transformers. AI Open
43. Liu X, Zhang F, Hou Z, Mian L, Wang Z, Zhang J (2021) Generative or contrastive. IEEE Trans Knowl Data Eng Self-superv Learn
44. Liu Y, Zhang Y, Wang Y, Hou F, Yuan J, Tian J, Zhang Y, Shi Z, Fan J, He Z (2021) A survey of visual transformers. arXiv:2111.06091
45. Liu Y, Ott M, Goyal N, Du J, Joshi M, Chen D, Levy O, Lewis M, Zettlemoyer K, Stoyanov V (2019) Roberta: a robustly optimized bert pretraining approach. arXiv:1907.11692
46. Mondal AK, Bhattacharjee A, Singla P, Prathosh AP (2021) xvitcos: explainable vision transformer based covid-19 screening using radiography. IEEE J Trans Eng Health Med 10:1–10
47. Otter DW, Medina JR, Kalita JK (2020) A survey of the usages of deep learning for natural language processing. IEEE Trans Neural Netw Learn Syst 32(2):604–624
48. Owais M, Lee YW, Mahmood T, Haider A, Sultan H, Park KR (2021) Multilevel deep-aggregated boosted network to recognize covid-19 infection from large-scale heterogeneous radiographic data. IEEE J Biomed Health Inf 25(6):1881–1891
49. Parmar N, Vaswani A, Uszkoreit J, Kaiser L, Shazeer N, Ku A, Tran D (2018) Image transformer. In: International conference on machine learning. PMLR, pp 4055–4064
50. Qiu X, Sun T, Yige X, Shao Y, Dai N, Huang X (2020) Pre-trained models for natural language processing: a survey. Sci China Technol Sci 63(10):1872–1897
51. Radford A, Narasimhan K, Salimans T, Sutskever I et al (2018) Improving language understanding by generative pre-training

52. Rajpurkar P, Irvin J, Zhu K, Yang B, Mehta H, Duan T, Ding D, Bagul A, Langlotz C, Shpanskaya K et al (2017) Chexnet: radiologist-level pneumonia detection on chest x-rays with deep learning. arXiv:1711.05225
53. Ramachandran P, Parmar N, Vaswani A, Bello I, Levskaya A, Shlens J (2019) Stand-alone self-attention in vision models. Adv Neural Inf Proc Syst 32
54. MrDM Reddy, MrSkM Basha, MrMC Hari, MrN Penchalaiah (2021) Dall-e: creating images from text. UGC Care Group I J 8(14):71–75
55. Rives A, Meier J, Sercu T, Goyal S, Lin Z, Liu J, Guo D, Ott M, Zitnick CL, Ma J et al (2021) Biological structure and function emerge from scaling unsupervised learning to 250 million protein sequences. Proc Natl Acad Sci 118(15):e2016239118
56. Sandler M, Howard A, Zhu M, Zhmoginov A, Chen L-C (2018) Mobilenetv2: inverted residuals and linear bottlenecks. In: Proceedings of the IEEE conference on computer vision and pattern recognition, pp 4510–4520
57. Schwaller P, Laino T, Gaudin T, Bolgar P, Hunter CA, Bekas C, Lee AA (2019) Molecular transformer: a model for uncertainty-calibrated chemical reaction prediction. ACS Central Sci 5(9):1572–1583
58. Simonyan K, Zisserman A (2014) Very deep convolutional networks for large-scale image recognition. arXiv:1409.1556
59. Su W, Zhu X, Cao Y, Li B, Lu L, Wei F, Dai J (2019) Vl-bert: pre-training of generic visual-linguistic representations. arXiv:1908.08530
60. Sun C, Shrivastava A, Singh S, Gupta A (2017) Revisiting unreasonable effectiveness of data in deep learning era. In: Proceedings of the IEEE international conference on computer vision, pp 843–852
61. Sun L, Mo Z, Yan F, Xia L, Shan F, Ding Z, Song B, Gao W, Shao W, Shi F et al (2020) Adaptive feature selection guided deep forest for covid-19 classification with chest ct. IEEE J Biomed Health Inf 24(10):2798–2805
62. Szegedy C, Liu W, Jia Y, Sermanet P, Reed S, Anguelov D, Erhan D, Vanhoucke V, Rabinovich A (2015) Going deeper with convolutions. In: Proceedings of the IEEE conference on computer vision and pattern recognition, pp 1–9
63. Tan M, Le Q (2019) Efficientnet: rethinking model scaling for convolutional neural networks. In: International conference on machine learning. PMLR, pp 6105–6114
64. Vaswani A, Ramachandran P, Srinivas A, Parmar N, Hechtman B, Shlens J (2021) Scaling local self-attention for parameter efficient visual backbones. In: Proceedings of the IEEE/CVF conference on computer vision and pattern Recognition, pp 12894–12904
65. Vaswani A, Shazeer N, Parmar N, Uszkoreit J, Jones L, Gomez AN, Kaiser Ł, Polosukhin I (2017) Attention is all you need. Adv Neural Inf Proc Syst 30
66. Vaz Rodrigues L, Martins Y, Guimarães C, de Santis M, Marques A, Barata F (2011) Anatomy for the bronchologist: a prospective study of the normal endobronchial anatomic variants. Revista Portuguesa de Pneumologia (English Edition) 17(5):211–215
67. Wang L, Lin ZQ, Wong A (2020) Covid-net: a tailored deep convolutional neural network design for detection of covid-19 cases from chest x-ray images. Sci Rep 10(1):1–12
68. Wang S, Li BZ, Khabsa M, Fang H, Ma H (2020) Linformer: self-attention with linear complexity. arXiv:2006.04768
69. Wang X, Girshick R, Gupta A, He K (2018) Non-local neural networks. In: Proceedings of the IEEE conference on computer vision and pattern recognition, pp 7794–7803
70. Wang Z, Liu Q, Dou Q (2020) Contrastive cross-site learning with redesigned net for covid-19 ct classification. IEEE J Biomed Health Inf 24(10):2806–2813
71. Woo S, Park J, Lee J-Y, Kweon IS (2018) Cbam: convolutional block attention module. In: Proceedings of the European conference on computer vision (ECCV), pp 3–19
72. Wu B, Xu C, Dai X, Wan A, Zhang P, Yan Z, Tomizuka M, Gonzalez J, Keutzer K, Vajda P (2020) Visual transformers: token-based image representation and processing for computer vision. arXiv:2006.03677
73. Xie X, Zhong Z, Zhao W, Zheng C, Wang F, Liu J (2020) Chest ct for typical coronavirus disease 2019 (covid-19) pneumonia: relationship to negative rt-pcr testing. Radiology 296:E41-e45

74. Yu X, Lu S, Lili G, Shui-Hua W, Yu-Dong Z (2021) Resgnet-c: a graph convolutional neural network for detection of covid-19. Neurocomputing 452:592–605
75. Zhu X, Su W, Lu L, Li B, Wang X, Dai J (2020) Deformable detr: deformable transformers for end-to-end object detection. arXiv:2010.04159
76. Zhuang L, Wayne L, Ya S, Jun Z (2021) A robustly optimized bert pre-training approach with post-training. In: Proceedings of the 20th Chinese national conference on computational linguistics, pp 1218–1227

COVID-19 Classification Using CT Scans with Convolutional Neural Networks

Pedro Crosara Motta, Paulo Cesar Cortez, and Jao Alexandre Lobo Marques

Abstract Even with more than 12 billion vaccine doses administered globally, the Covid-19 pandemic has caused several global economic, social, environmental, and healthcare impacts. Computer Aided Diagnostic (CAD) systems can serve as a complementary method to aid doctors in identifying regions of interest in images and help detect diseases. In addition, these systems can help doctors analyze the status of the disease and check for their progress or regression. To analyze the viability of using CNNs for differentiating Covid-19 CT positive images from Covid-19 CT negative images, we used a dataset collected by Union Hospital (HUST-UH) and Liyuan Hospital (HUST-LH) and made available at the Kaggle platform. The main objective of this chapter is to present results from applying two state-of-the-art CNNs on a Covid-19 CT Scan images database to evaluate the possibility of differentiating images with imaging features associated with Covid-19 pneumonia from images with imaging features irrelevant to Covid-19 pneumonia. Two pre-trained neural networks, ResNet50 and MobileNet, were fine-tuned for the datasets under analysis. Both CNNs obtained promising results, with the ResNet50 network achieving a Precision of 0.97, a Recall of 0.96, an F1-score of 0.96, and 39 false negatives. The MobileNet classifier obtained a Precision of 0.94, a Recall of 0.94, an F1-score of 0.94, and a total of 20 false negatives.

P. C. Motta · P. Cesar Cortez
Laboratório de Engenharia e Sistemas de Computação, Universidade Federal do Ceará, Campus do Pici, Fortaleza, Brazil
e-mail: pedro.motta@lesc.ufc.br

P. Cesar Cortez
e-mail: cortez@lesc.ufc.br

J. A. Lobo Marques (✉)
Laboratory of Applied Neurosciences, University of Saint Joseph, Estrada Marginal da Ilha Verde, 14-17 Macao SAR, China
e-mail: alexandre.lobo@usj.edu.mo

J. A. Lobo Marques and S. J. Fong (eds.), *Computerized Systems for Diagnosis and Treatment of COVID-19*, https://doi.org/10.1007/978-3-031-30788-1_7

1 Introduction

Even with more than 12 billion vaccines doses administered globally, the Covid-19 pandemic has caused several global impacts in economic, social, environmental, and healthcare areas [3, 14, 20].

An early diagnosis of the disease is crucial for treatment and rapid patient isolation, which may contain virus propagation. Furthermore, it is necessary to optimize resource utilization for treatment, as Covid-19 treatment can be long [24].

Computer Aided Diagnostic (CAD) systems can serve as a complementary method to aid doctors in identifying regions of interest in images and help detect diseases. In addition, these systems can help doctors analyze the status of the disease and check for their progress or regress [25].

Medical imaging systems, such as computerized tomography (CT) scans, are crucial in disease diagnosis and follow-up because they are non-invasive and have high sensibility. Despite the overwhelming two trillion images produced annually worldwide, the world is facing a 12.9 million medical professionals shortage by 2035. Moreover, in many countries, such as the United States, the United Kingdom, and China, the number of radiologists is insufficient [13].

Chest CT scans have been used to train CNNs to diagnose lung diseases such as lung cancer, tuberculosis, chronic obstructive pulmonary disease (COPD), pneumonia (COVID-19), pulmonary embolism, and interstitial lung disease (ILD). Specialists label Chest CT scans datasets in order to train CNNs. These labels can be for classification, detection, or segmentation tasks [2, 7, 10, 13, 18, 26, 28].

Computed tomography (CT) of the chest is an efficient way to diagnose suspected SARS-CoV-2 patients, as pneumonia caused by COVID-19 is one of the most common clinical presentations of COVID-19 [32].

In the CT analysis, the main characteristics present in patients with COVID-19 are: ground-glass opacity (88.0%), bilateral involvement (87.5%), peripheral distribution (76.0%), and multilobar involvement (78.8%) [21].

Ground-glass opacity represents the partial filling of the alveoli (or partial collapse of the alveoli), increased blood supply, or both together. On CT images, ground-glass opacity appears as an increase in lung parenchyma density.

The main objective of this chapter is to present results from applying two state-of-the-art CNNs on a Covid-19 CT image dataset to evaluate the possibility of differentiating images with imaging features associated with Covid-19 pneumonia from images with imaging features irrelevant to Covid-19 pneumonia.

2 Related Works

The work presented in [28] proposes a CNN for the task defined as PN-SAMP (Pulmonary Nodule Segmentation Attributes and Malignancy Prediction). It can accurately predict the malignancy of lung nodules and provide high-level semantic attributes and the areas of detected nodules. Moreover, the combination of nodule segmentation, attributes, and malignancy prediction helps improve the performance of every single task. In addition, inspired by the fact that radiologists often change window widths and window centers to help to decide on uncertain nodules, PN-SAMP mixes multiple WW/WC to gain information for the raw CT input images. The system evaluation uses the public dataset LIDC-IDRI, which is one of the largest datasets for lung nodule malignancy prediction. The results presented after running the proposed PN-SAMP achieved accuracies of 97.58% for nodule malignancy prediction, 89.33% for attributes prediction, and finally, a Dice Coefficient of 73.89% for nodule segmentation. The performance is significant for different tasks of segmentation and classification of lung nodules, in addition to the possibility of attribute learning.

The application of deep learning to computerized chest tomography (CT) to assist doctors in detecting and differentiating Background Pulmonary tuberculosis (PTB) from non-tuberculous (non-TB) pneumonia in an expedient, non-invasive and reproducible manner is proposed in [26]. They collected a dataset containing 1,124 CT scans from 923 PTB and non-TB pneumonia based on their pathological reports of lung biopsy and clinical information and 201 patients without pulmonary infiltrate from West China Hospital between 2012 and 2018. Randomly selected parts of this dataset (WCPR dataset) were used to develop, train, internally validate, and test the algorithm. Patients in the WCPR dataset (PTB, n = 439; non-TB pneumonia, n = 484; normal, n = 201) were randomly assigned to three non-overlapping sets: training set, n = 866; validation set, n = 108; and test set, n = 150. An additional dataset from NIH TB Portal31 comprising cases from three countries (Belarus, n = 274; Romania, n = 43; Moldova, n = 10) was used to validate externally the algorithm's ability to identify PTB. A convolutional neural network of Inception-Res-Net-v230 was trained and tested on the entire chest CT to mimic real-life applications. The performance of our algorithm was compared with three trained radiology/pulmonology physicians. When differentiating pulmonary infiltrates, the algorithm achieved 99.3% accuracy (149 out of 150), 99.0% sensitivity, and 100.0% specificity. For identifying PTB, the algorithm achieved 82.0% accuracy (123 out of 150), 95.9% sensitivity, and 75.2% specificity. For identifying non-TB pneumonia, the algorithm achieved 81.3% accuracy (122 out of 150), 52.7% sensitivity, and 97.9% specificity. These results mostly outperformed our human readers for PTB identification, who averaged up to 81.1% accuracy, 70.8% sensitivity, and 86.1% specificity. In addition, our algorithm identified 287 out of 327 PTB (87.8% accuracy) cases in NIH TB Portal Dataset from other countries.

In another work, a new classification method for Chronic obstructive pulmonary disease (COPD) grouping is proposed, considering the use of Deep Learning and another technique called PRM—Parametric-Response mapping method. The authors proposed as input parameters of the 3D convolutional neural network (CNN), an image created from the extraction of parenchymal functional variables of functional small airway disease percentage (fSAD%) and emphysema percentage (Emph%) [7]. The results for the classification are an accuracy of 89.3% and a sensitivity of 88.3%, considering a five-fold cross-validation for the model integrating the 3D-

CNN and PRM (3D-cPRM). In addition, the accuracy of the prediction task of the new model exceeded the results from the literature, which are based on 2D and 3D CNNs models, even with the same neural network. Following this, they applied a gradient-weighted class activation mapping (Grad-CAM), which is an Explainable AI (XAI) technique, highlighting the main features of the CNN learning process. The technique shows that most of the class-discriminative regions appeared in the upper and middle lobes of the lung, which is consistent with the regions of elevated fSAD% and Emph% in COPD subjects. As a main result, it was possible to use the 3D-cPRM model to determine parenchymal abnormalities in patients with COPD with a significant correlation with the diagnostics based on CT-Scans.

Because of the rapid manifestations of COVID-19 and the significant number of disease cases, Artificial Intelligence (AI) studies were developed to aid medical diagnosis in computed tomography (CT) images.

The authors in [11] developed a 2D structure called fast-track COVID-19 classification network (FCONet) to diagnose COVID-19 pneumonia based on a chest CT image. They developed FCONet using VGG16, ResNet-50, Inception-v3, and Xception transfer learning networks as a backbone. They used 3,993 chest CT images of patients with COVID-19 pneumonia, other types of pneumonia, and non-pneumonic diseases from Wonkwang University Hospital (WKUH), Chonnam National University Hospital (CNUH), and the Society's public database), and from the public database of Italian Society of Medical and Interventional Radiology (SIRM). Images were resized to 256x256 pixels and normalized to pixel values between 0 and 1. Then, they applied data augmentation and 5-fold cross-validation. Among the four pre-trained FCONet models, ResNet-50 presented the best performance, with rates of sensitivity of 99.58%, specificity of 100.00%, and accuracy of 99.87%. Authors evaluated FCONet models on an external test dataset [30]; the detection accuracy rates of the ResNet-50 model stood out as the highest at 96.97%, followed by Xception, Inception-v3, and VGG16 with values equal to 90.71 %, 89.38%, and 87.12%, respectively.

A self-developed model called CTnet-10 for the diagnosis of COVID-19 is designed in [23]. The authors also compared the proposed model with DenseNet-169, VGG-16, ResNet-50, InceptionV3, and VGG-19. They used the CT COVID-19 dataset, which consists of 738 CT images, with 349 images from 216 patients having COVID-19, while 463 images were from non-COVID-19 patients [30]. The images were divided into training, validation, and test subsets with a division of 80%, 10%, and 10%, respectively. The self-developed CTnet-10 model presented the shortest training, testing, and execution time, reaching an accuracy rate of 82.10%. However, VGG-19 was superior, with an accuracy of 94.52% compared to all other deep learning models.

The research in [12] presented a method that combines Stacked Generalization ensemble learning with VGG16 deep learning to form a cascade classifier. The dataset used was the CC-CCII [31]. The training set is divided into six sub-training groups and a triclassifier, and five binary classifiers are trained. The training and testing of the classification model are made only on 2D scan images instead of full CT scans. The original chest CT image size of 512×512 is kept. The algorithm has been validated to classify patients with COVID-19, common pneumonia (CP), and normal controls. The results achieved an accuracy of 93.57%, a sensitivity of 94.21%, a specificity of 93.93%, a precision of 89.40%, and an F1-score of 91.74% for the three categories.

In [33], the authors used an ensemble method denoted by EDL-COVID. AlexNet, GoogleNet, and ResNet were used to extract features from images. Data were obtained from previous publications, official media reports, and public databases [29]. CT images were preprocessed, excluding images with high noise and coronal position, and all images were normalized and resized to 64 × 64. The final dataset contains 2500 images of each class: COVID-19, lung tumor, and normal lung. Softmax was used in the fully connected layer classification algorithm. The result with EDL-COVID achieved a 5-fold average accuracy of 99.05%, a sensitivity of 99.05%, a specificity of 99.60%, an F-score 98.59%, and Matthews correlation coefficient (MCC) of 97.89%.

In addition, a feature-wise attention layer to enhance the discriminative features obtained by convolutional networks is proposed in [17]. The dataset used was COVID-CT [30]. All images were resized to 256 × 256, and pixel values were normalized between 0 and 1. The authors applied the mixup technique of data augmentation to improve the network performance. ResNet50 architecture was extended with a resource attention layer to achieve an accuracy of 95.57%, f1-score: 95.61%, and AUC of 95.71%.

Wang proposed a backbone redesigning COVID-Net for CT images [27]. Previously this model was applied for X-ray images. A joint learning scheme was applied to address the data heterogeneity problem, evaluating two datasets, SARS-CoV-2 and COVID-CT. The images were resized to 224x224 and normalized with zero mean and unit variance. 4-fold cross-validation was used to generalize the results. The best results were with the SARS-CoV-2 dataset, achieving an average accuracy rate, f1-score, sensitivity, precision, and AUC of 90.83%, 90.87%, 85.89%, 95.75%, and 96.24%, respectively.

The authors applied a CNN-based network with the SqueezeNet model to distinguish COVID-19 CT images from other pneumonia and healthy CT images in [19]. Two COVID-CT datasets and the SIRM were used. The first comprises 360 CT scans of individuals with COVID-19 and 397 of different disease types and healthy individuals. The Italian dataset is composed of 100 CT scans of COVID-19. In both datasets, a data augmentation was performed on the training data. 10-fold cross-validation was used. The results obtained outperformed the original SqueezeNet with the CNN-2 network trained and validated on dataset arrangement 2, achieving an accuracy of 85.03%, a sensitivity of 87.55%, a specificity of 81.95%, a precision of 85.01%, and an f1-score of 86.20%. After the overview of related works provided in this Section, the proposed methodology is presented in the following Section.

3 Materials and Methods

This section describes in detail the workflow for the proposed methodology of this chapter. We used preprocessed images with extracted lung parenchyma, randomly undersampled the imbalanced negative Covid-19 class, fine-tuned pre-trained ResNet50 and MobileNet with this data, and then calculated evaluation metrics. The workflow flowchart is shown in Fig. 1.

3.1 Dataset

In order to analyze the viability of using CNNs for differentiating Covid-19 CT positive images from Covid-19 CT negative images, we used a dataset collected by Union Hospital (HUST-UH) and Liyuan Hospital (HUST-LH) [16] and made available at Kaggle platform [1].

Authors manually labeled 19,685 CT slices exported from DICOM images after removing personal information for 61 COVID-19 patients and 43 control cases (Fig. 2).

Individual CT images in JPEG format were classified into three types: **(I)** 5705 non-informative CT (NiCT) images where lung parenchyma was not captured for any judgment, **(II)** 4001 positive CT (pCT) images where imaging features associated with COVID-19 pneumonia could be unambiguously discerned, and **(III)** 9979 negative CT (nCT) images where imaging features in both lungs were irrelevant to COVID-19 pneumonia. It is essential to point out that images of nCT type might have anomalies other than Covid-19 (Fig. 3).

The authors provided preprocessed images with extracted lung parenchyma. Therefore, we used preprocessed images from the pCT and nCT classes in this chapter. Furthermore, we undersample the nCT class to balance the data by randomly selecting 4001 nCT images.

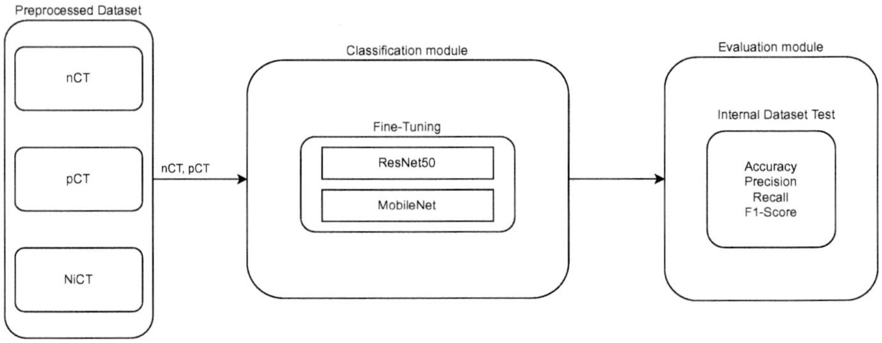

Fig. 1 Methodology flowchart for this chapter

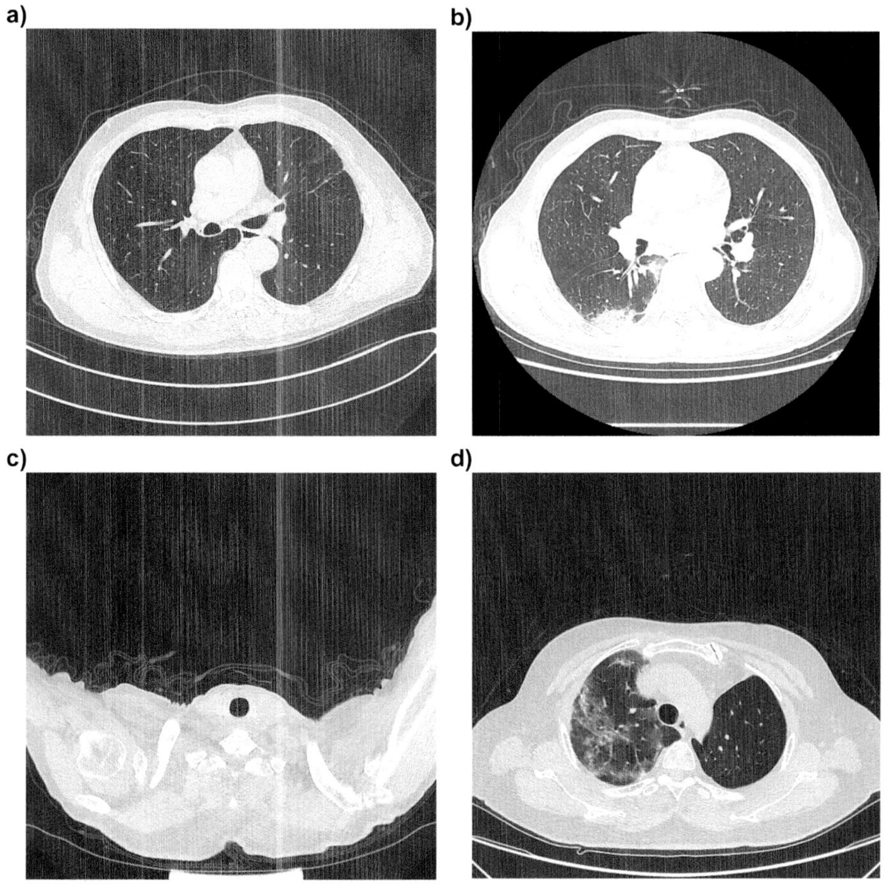

Fig. 2 Presentation of CT-Scan Samples from the selected dataset. **a** negative CT (nCT) image without anomalies. **b** negative CT (nCT) image with anomalies on the left lung. **c** non-informative CT (NiCT) image. **a** positive CT (pCT) image

Finally, we divided our data into train/validation/test splits to assess our models. We used 5616 images for training, 625 for validation, and 1761 for testing.

In order to prevent overfitting, we performed data augmentation with random horizontal flips, random rotations in the angle range [–15, 15], and randomly applied a Gaussian blur.

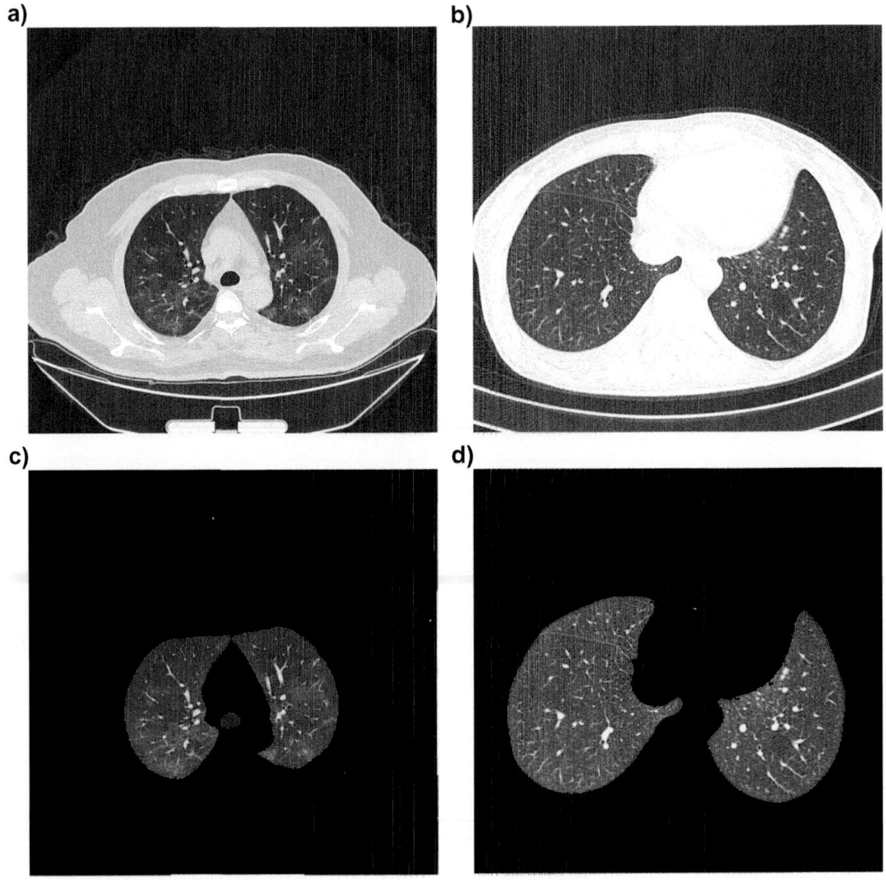

Fig. 3 Samples from the used dataset. **a** positive CT (pCT) image. **b** negative CT (nCT) image. **c** Preprocessed pCT image. **d** Preprocessed nCT image

3.2 CNN Models for COVID-19 Classification

For the proposed solution, two different transfer learning models will be compared. One is a pre-trained ResNet50, and the other is a pre-trained MobileNet.

Both were fine-tuned for working with our datasets. The adjustment of pre-trained deep learning models is a necessary step to allow better classification during new classification tests they are submitted to. Within this step, the pre-trained fully-connected layer from ImageNet is removed. The modeling approach is represented in Fig. 4.

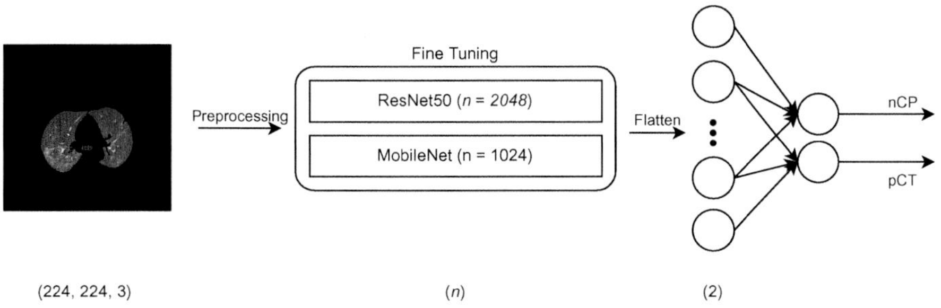

Fig. 4 Architecture of the classification model using ResNet50 and MobileNet Fine-Tuned CNNs

From the model presented, it is possible to see that a new fully connected layer with two output neurons is added. The two output neurons represent the classification of the CT Scan image under analysis, indicating if it's a positive or a negative classification result for COVID-19.

3.2.1 MobileNet

Google researchers proposed MobileNet architecture to fit mobile applications. Mobilenet is based on the theory of depthwise separable convolutions, which makes the architecture simple, reducing the need for high computational power. Depth convolution and point convolution replace the standard convolutional filters to create a detachable depth filter.

The core layer of Mobilenet, which presents one of the essential properties of this network, is the separable convolutions in depth. The main feature of depth convolutions is to split the standard convolution into two nuclei, a depth convolution and a point convolution 1×1. As a result, deep-separable convolution uses 8-9 times less computational cost than standard convolution [8].

3.2.2 ResNet50

Researchers at Microsoft Research developed residual neural networks (ResNet). The central aspect differentiating it from other CNN architectures is the residual block. This block uses shortcuts between layers, adding the initial values of the layer's input to the ReLU output function $y = F(x, W_i)) + x$ of the layer, where the function $F(x, W_i)$ represents the residual map learned by the block and x the initial image.

The layers in a traditional CNN are learning the actual $(H(x))$ output, while residual network layers are learning the residual $(R(x))$. These shortcut connections skip over layers and are arranged in residual blocks. These blocks have three convolution layers with k filters 1×1, 3×3, and 1×1, respectively, for each layer. Given an input x, the residual mapping $F(x)$ is denoted by $F(x) + x$. This mapping adds the input of a residual block (x identity) to the output resulting from that same block [6].

A recent publication [4] considered both models for mammography analysis, which is considered the gold standard for screening for breast abnormalities since it facilitates early detection and diagnosis of disease. Breast cancer accounts for a significant incidence of female mortality worldwide. The work considered the use of deep learning models. The authors presented results on malignancy classification of breast abnormalities using transfer learning, including MobileNet and NasNet (which is not considered in this Chapter), and their performance is compared with Resnet50. Resnet50 and MobileNet achieved the best performance, respectively with 78.4 and 74.3%, of Accuracies.

3.3 *Evaluation Metrics*

Both models were evaluated on test data using the following metrics:

- Accuracy—which is a measure considering the overall performance of the classifier to detect True Positives (TP) and True Negatives (TN):

$$Accuracy = \frac{TP + TN}{TP + TN + FP + FN} \quad (1)$$

- Precision—determining the classifier performance to detect True Positive (TP) cases, according to the False Positives (FP):

$$P = \frac{TP}{TP + FP} \quad (2)$$

- Recall—determining the classifier performance to detect True Positive (TP) cases, according to the False Negatives (FN):

$$R = \frac{TP}{TP + FN} \quad (3)$$

- F1-Score—which is a metric based on the two previous metrics, Precision and Recall, with the aim to indicate a performance index considering False Positives (FP) and False Negatives (FN):

$$F1 = \frac{2 * (Precision * Recall)}{Precision + Recall} \quad (4)$$

3.4 Explainable AI (XAI) Techniques

Machine Learning techniques such as Neural Networks or Forest-based classifiers may generate complex mathematical models with multiple layers and interconnections, which might become hard to identify when which steps of the intended process are performed. For example, analyzing one medical image and providing support for diagnostic may raise several questions, such as: in which part of the network is the image segmented? When/where the attributes for classification are extracted? When/where is the classification performed? This characteristic of these complex structures created significant credibility issues since the steps would not be easily determined, and classification bias could be embedded in the model, even inadvertently.

Artificial Intelligence tends to become ubiquitous in medical diagnostic support systems. This affirmation does not mean that the systems will substitute the specialists or perform the complete set of classifications in the short term. But for the task of processing complex medical imaging from different sources, AI systems can perform additional tasks very well, such as image filtering, noise reduction or removal, recovering image definition, segmentation of specific areas, classification of specific diseases, etc. Nevertheless, all these tasks should be trustworthy and open for human auditing.

Integrating machines and algorithms into our daily lives require explainability to increase social acceptance. The results achieved by a proposed model are important, but understanding the process of achieving these results is the core of the necessary credibility. Medical specialists demand a high-level of explainability to be able to trust in a computerized medical diagnostic supporting system [9, 15].

A new concept called XAI—Explainable Artificial Intelligence was proposed to respond to these challenges. The idea is to develop computational models that can clearly present the results of deep learning and machine learning systems for humans to understand and trust. It should be able to detect bias and validate the proposed approach, according to its goals, for example, if it is a segmentation, a classification, or any other task.

Explainability is the degree to which a human can understand the cause of a decision made by a model. The higher the explainability of a machine learning model, the easier it will be for someone to comprehend why that specific model makes certain decisions or predictions. Explainability is helpful as a debugging tool for detecting bias [15].

Understanding and explaining not only "How" but also "Why" a proposed deep learning network achieved one specific result is essential. The capacity of explanation will allow systems external quality certification since external auditors may be able to verify its unique strengths and the impact on clinical processes. In addition, with user-friendly interfaces for the XAI tools, it becomes possible to implement systems improvement processes since teams of specialists may be able to provide insights and raise issues following the explainable reports.

With the objective of presenting an overview of this growing approach in Artificial Intelligence, four relevant XAI techniques are presented in this Section:

- LIME
- Smooth-GRAD
- CAM
- GradCam

3.4.1 LIME

The first technique for image processing is based on analyzing the changes in one model response following a change in the original image, which can be editing, modifying, or adding noise. It is called LIME—Locally Interpretable Model Explanations.

Analyzing how perturbations in the input data affect the final prediction determines which part of the original image may be relevant for an eventual classification task [9]. An example is shown in Fig. 5.

3.4.2 Smooth-Grad

For gradient-based methods, such as Smooth-Grad, the gradient dy/dx between the output model class Y and the input image X is used to calculate saliency maps. For large networks, the gradient information is very noisy, so there have been many modifications to this method that reduce noise by smoothing, thresholding or rescaling [9].

Fig. 5 GramCam results for MobileNet. **a** Original Image. **b** LIME output for "bagel" class. *Source* https://christophm.github.io/interpretable-ml-book/. License: Creative Commons

3.4.3 CAM

The second technique is based on the concept of activation maps, such as CAM—Class Activation Mapping. To understand these maps, it is important to understand that deep learning networks have different types of layers, performing different sub-tasks, such as pooling or softmax layers. The focus here is to create feature maps and analyze if that map is relevant for the spatial analysis of the image. With a vast number of maps, it is possible to evaluate through exploratory testing if these features are relevant for the classification or the prediction task. [9].

3.4.4 GradCam

The last technique presented here is GradCam, a specific type of CAM (see previous Section). It follows a visual strategy to explain the functioning of a significant number of Deep Learning CNN (Convolutional Neural Networks) models.

Consider, for example, a network that should identify one concept, such as specific objects in a photo, like a person or a robot. As presented in the technique name, it considers a gradient through the layers of the classifier network, with the aim of creating a map clearly identifying relevant regions of the original image, as shown in Fig. 6. As presented in the literature, these gradient based CAM explainable technique may be used in different CNN structures, such as fully-connected layers networks, reinforcement learning, networks with multi-modal inputs, and many others [22].

3.5 XAI Implementation

In this Chapter, to better understand our models' decisions, we used Gradient-weighted Class Activation Mapping (GradCam), an explainable technique to better evaluate the trained models [5].

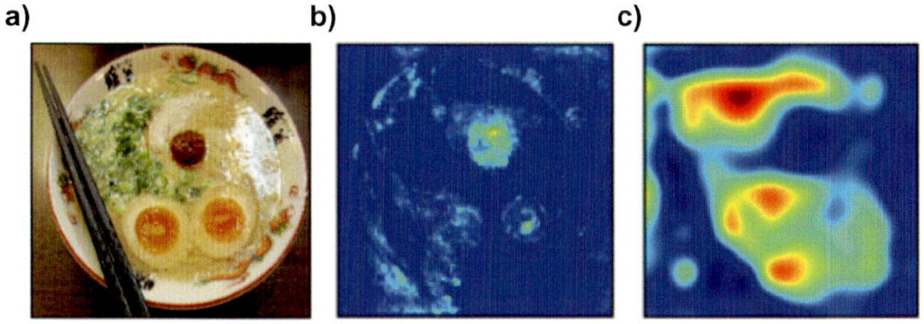

Fig. 6 GramCam results for MobileNet. **a** Original image. **b** Smooth-Grad output for "soup bowl" class. **c** GradCam output for "soup bowl" class. *Source* https://christophm.github.io/interpretable-ml-book/ License: Creative Commons

4 Results

Both CNNs obtained promising results, with the ResNet50 model achieving a Precision of 0.97, a Recall of 0.96, an F1-score of 0.96, while the MobileNet model obtaining a precision of 0.94, a recall of 0.94, and an F1-score of 0.94 (Tables 1 and 2).

However, even if metrics from ResNet50 appear to be better, when we compare Confusion Matrices (Figs. 7 and 9) from both models, it is possible to see that MobileNet achieved a lower number of False Negatives—20 false negatives—when compared with the ResNet50—39 false negatives.

From a clinical perspective, it can be assumed to be more harmful or risky to have a false negative (not detecting a positive case in the CT Scan analyses) than to have a false positive (erroneously detecting a negative case as positive), MobileNet might be the safest model in this analysis.

GradCam results are heatmaps; the higher the color intensity, the higher the region's importance to the model decision. For example, we can see in Fig. 8 that the most important regions in the lung image coincide with the Covid-19 lesions. These results might show that our ResNet50 model correctly analyzes our images, "paying more attention" to the correct regions.

For MobileNet (Fig. 10), the region detected by GradCam coincides with the Covid-19 lesions in the right lung. However, a small lesion in the left lung is not "looked at."

Table 1 ResNet50 classification results

	Precision	Recall	F1-score	Support
nCP	0.96	0.97	0.97	894
pCP	0.97	0.96	0.96	867
Accuracy			0.96	1761
Macro avg	0.97	0.96	0.96	1761
Weighted avg	0.96	0.96	0.96	1761

Table 2 MobileNet classification results

	Precision	Recall	F1-score	Support
nCP	0.98	0.91	0.94	894
pCP	0.91	0.98	0.94	867
Accuracy			0.94	1761
Macro avg	0.94	0.94	0.94	1761
Weighted avg	0.94	0.94	0.94	1761

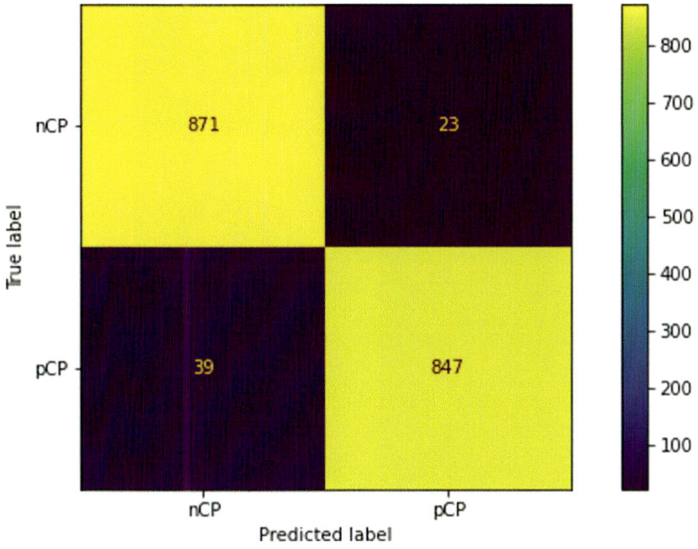

Fig. 7 ResNet50 confusion matrix

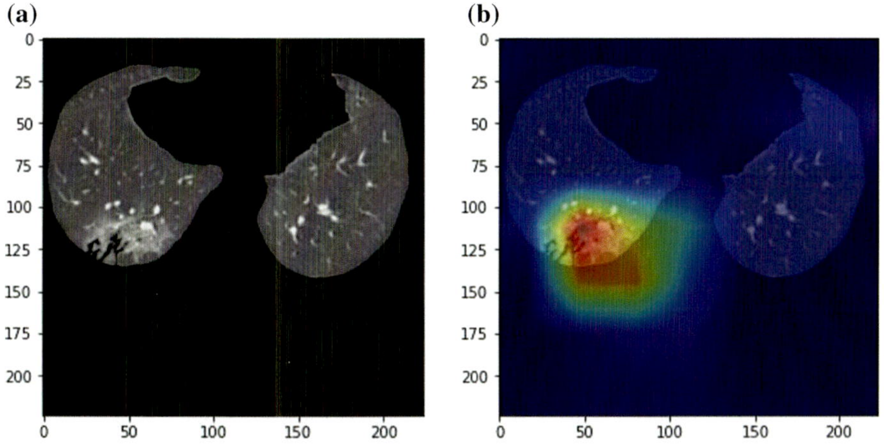

Fig. 8 GramCam results for ResNet50. **a** Original image. **b** GradCam output

5 Conclusion

The efficient computerized analysis of CT-Scans images for the classification of pneumonia that may indicate COVID-19 or other pulmonary diseases is a complex task, which can be used to benefit the medical decision-making process for diagnosis or patient management.

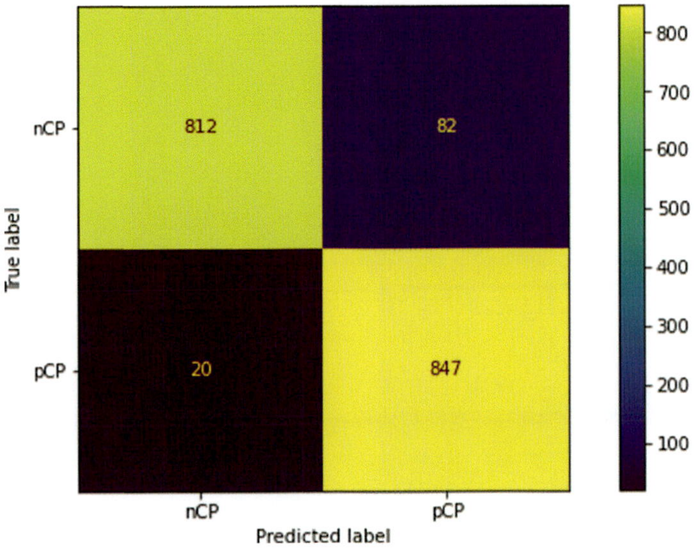

Fig. 9 MobileNet confusion matrix

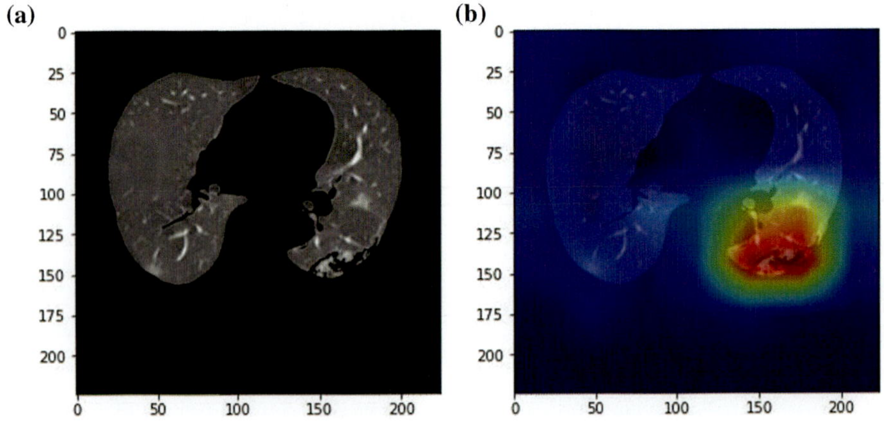

Fig. 10 GramCam results for MobileNet. **a** Original image. **b** GradCam output

In this Chapter, we used ResNet50 and MobileNet architecture to classify lung CT images with imaging features associated with Covid-19 pneumonia from images with imaging features irrelevant to Covid-19 pneumonia. Both networks obtained promising results, with the ResNet50 model achieving a Precision of 0.97, a Recall of 0.96, and F1-score of 0.96, while the MobileNet model obtaining a precision of 0.94, a recall of 0.94, and an F1-score of 0.94. From the perspective of false negatives, the MobileNet achieved a total of 20 false negatives, while the ResNet50 failed to classify 39 positive cases.

Future works may compare the proposed models with other convolutional network models and use further XAI techniques to comprehend better the system performance, benefits and limitations.

References

1. Ct scans for covid-19 classification — kaggle. https://www.kaggle.com/datasets/azaemon/preprocessed-ct-scans-for-covid19. (Accessed 13 Nov 2022)
2. Chassagnon G, Vakalopoulou M, Régent A, Sahasrabudhe M, Marini R, Hoang-Thi T-N, Dinh-Xuan A-T, Dunogué B, Mouthon L, Paragios N, Revel M-P (2021) Elastic registration-driven deep learning for longitudinal assessment of systemic sclerosis interstitial lung disease at ct. Radiology 298(1):189–198. PMID: 33078999
3. Gardner L, Dong E, Du H (2020) An interactive web-based dashboard to track covid-19 in real time. Lancet Inf Dis 20(5):533–534
4. Falconí LG, Pérez M, Aguilar WG (20198) Transfer learning in breast mammogram abnormalities classification with mobilenet and nasnet. In: 2019 international conference on systems, signals and image processing (IWSSIP), pp 109–114
5. Gildenblat J, and Contributors (2021) Pytorch library for cam methods. https://github.com/jacobgil/pytorch-grad-cam
6. He K, Zhang X, Ren S, Sun J (2016) Deep residual learning for image recognition. In: 2016 IEEE conference on computer vision and pattern recognition (CVPR), pp 770–778
7. Ho TT, Kim T, Kim WJ, Lee CH, Chae KJ, Bak SH, Kwon SO, Jin GY, Park EK, Choi S (2021) A 3d-cnn model with ct-based parametric response mapping for classifying copd subjects. Sci Rep 11(1):34
8. Howard AG, Zhu M, Chen B, Kalenichenko D, Wang W, Weyand T, Andreetto M, Adam H (2017) Mobilenets: efficient convolutional neural networks for mobile vision applications. CoRR arXiv:abs/1704.04861
9. Hryniewska W, Bombiński P, Szatkowski P, Tomaszewska P, Przelaskowski A, Biecek P (2021) Checklist for responsible deep learning modeling of medical images based on covid-19 detection studies. Pattern Recognit 118:108035
10. Huang SC, Kothari T, Banerjee I, Chute C, Ball RL, Borus N, Huang A, Patel BN, Rajpurkar P, Irvin J, Dunnmon J, Bledsoe J, Shpanskaya K, Dhaliwal A, Zamanian R, Ng AY, Lungren MP (2020) Penet—a scalable deep-learning model for automated diagnosis of pulmonary embolism using volumetric ct imaging. npj Digit Med 3(1):61
11. Ko H, Chung H, Kang WS, Kim KW, Shin Y, Kang SJ, Lee JH, Kim YJ, Kim NY, Jung H, Lee J et al (2020) Covid-19 pneumonia diagnosis using a simple 2d deep learning framework with a single chest ct image: model development and validation. J Med Internet Res 22(6):e19569
12. Li X, Tan W, Liu P, Zhou Q, Yang J (2021) Classification of covid-19 chest ct images based on ensemble deep learning. J Healthcare Eng 2021
13. Liu F, Tang J, Ma J, Wang C, Ha Q, Yizhou Yu, Zhou Z (2021) The application of artificial intelligence to chest medical image analysis. Intell Med 1(3):104–117
14. Mattioli AV, Ballerini Puviani M, Nasi M, Farinetti A (2020) Covid-19 pandemic: the effects of quarantine on cardiovascular risk. Europ J Clin Nutrition 74(6):852–855
15. Molnar C (2022) Interpretable machine learning, 2 edn
16. Ning W, Lei S, Yang J, Cao Y, Jiang P, Yang Q, Zhang J, Wang X, Chen F, Geng Z, Xiong L, Hongmei Z, Yaping G, Yulan Z, Heshui S, Wang L, Yu X, Wang Z (2020) Open resource of clinical data from patients with pneumonia for the prediction of covid-19 outcomes via deep learning. Nature Biomed Eng 4(12):1197–1207
17. Özdemir Ö, Sönmez EB (2022) Attention mechanism and mixup data augmentation for classification of covid-19 computed tomography images. J King Saud Univers-Comput Inf Sci 34(8):6199–6207

18. Pathak Y, Shukla PK, Tiwari A, Stalin S, Singh S, Shukla PK (2022) Deep transfer learning based classification model for covid-19 disease. IRBM 43(2):87–92
19. Polsinelli M, Cinque L, Placidi G (2020) A light cnn for detecting covid-19 from ct scans of the chest. Pattern Recogn Lett 140:95–100
20. Rume T, Didar-Ul Islam SM (2020) Environmental effects of covid-19 pandemic and potential strategies of sustainability. Heliyon 6(9):e04965
21. Salehi S, Abedi A, Balakrishnan S, Gholamrezanezhad A (2020) Coronavirus disease 2019 (covid-19): a systematic review of imaging findings in 919 patients. Am J Roentgenol 215(1):87–93. PMID: 32174129
22. Selvaraju RR, Cogswell M, Das A, Vedantam R, Parikh D, Batra D (2017) Grad-cam: Visual explanations from deep networks via gradient-based localization. In: Proceedings of the IEEE international conference on computer vision, pp 618–626
23. Shah V, Keniya R, Shridharani A, Punjabi M, Shah J, Mehendale N (2021) Diagnosis of covid-19 using ct scan images and deep learning techniques. Emerg Radiol 28(3):497–505
24. Soares E, Angelov P, Biaso S, Froes MH, Abe DK (2020) Sars-cov-2 ct-scan dataset: a large dataset of real patients ct scans for sars-cov-2 identification. medRxiv
25. Valente IRS, Cortez PC, Neto EC, Soares JM, de Albuquerque VHC, Tavares JMRS (2016) Automatic 3d pulmonary nodule detection in ct images: a survey. Comput Methods Programs Biomed 124:91–107
26. Wang Y, Mei X, Liu C, Deyer T, Zeng J, Xia C, Schefflein J, Jia L, Yu H, Jiang F, Yang C, Zhou P, Chang H, Robson P, Doshi A, Mendelson D, Zhu H, Powell C, Yang Y, Li W (2019) A generalized deep learning approach for evaluating secondary pulmonary tuberculosis on chest computed tomography. SSRN Electron J 01
27. Wang Z, Liu Q, Dou Q (2020) Contrastive cross-site learning with redesigned net for covid-19 ct classification. IEEE J Biomed Health Inform 24(10):2806–2813
28. Wu B, Zhou Z, Wang J, Wang Y (2018) Joint learning for pulmonary nodule segmentation, attributes and malignancy prediction. In: 2018 IEEE 15th international symposium on biomedical imaging (ISBI 2018), pp 1109–1113
29. Xiong L, Zhong S, Li Y, Dai S, Bi L, Bo H, Zhang H, Zhimin M, Zeng S, Qi W et al (2020) Data-driven fuzzy multiple criteria decision making and its potential applications. Math Probl Eng 1(2):3
30. Yang X, He X, Zhao J, Zhang Y, Zhang S, Xie P (2020) Covid-ct-dataset: a ct scan dataset about covid-19
31. Zhang K, Liu X, Shen J, Li Z, Sang Y, Wu X, Zha Y, Liang W, Wang C, Wang K, Ye L, Gao M, Zhou Z, Li L, Wang J, Yang Z, Cai H, Xu J, Yang L, Cai W, Xu W, Wu S, Zhang W, Jiang S, Zheng L, Zhang X, Wang L, Lu L, Li J, Yin H, Wang W, Li O, Zhang C, Liang L, Wu T, Deng R, Wei K, Zhou Y, Chen T, Yiu-Nam Lau J, Fok M, He J, Lin T, Li W, Wang G (2020) Clinically applicable ai system for accurate diagnosis, quantitative measurements, and prognosis of covid-19 pneumonia using computed tomography. Cell 181(6):1423–1433.e11. 32416069[pmid]
32. Zhao W, Zhong Z, Xie X, Qizhi Yu, Liu J (2020) Relation between chest ct findings and clinical conditions of coronavirus disease (covid-19) pneumonia: a multicenter study. Am J Roentgenol 214(5):1072–1077. PMID: 32125873
33. Zhou T, Huiling L, Yang Z, Qiu S, Huo B, Dong Y (2021) The ensemble deep learning model for novel covid-19 on ct images. Appl Soft Comput 98:106885

TPOT Automated Machine Learning Approach for Multiple Diagnostic Classification of Lung Radiography and Feature Extraction

Francisco Nauber Bernardo Gois, Joao Alexandre Lobo Marques, and Simon James Fong

Abstract This chapter describes an AUTO-ML strategy to detect COVID on chest X-rays utilizing Transfer Learning feature extraction and the AutoML TPOT framework in order to identify lung illnesses (such as COVID or pneumonia). MobileNet is a lightweight network that uses depthwise separable convolution to deepen the network while decreasing parameters and computation. AutoML is a revolutionary concept of automated machine learning (AML) that automates the process of building an ML pipeline inside a constrained computing framework. The term "AutoML" can mean a number of different things depending on context. AutoML has risen to prominence in both the business world and the academic community thanks to the ever-increasing capabilities of modern computers. Python Optimised ML Pipeline (TPOT) is a Python-based ML tool that optimizes pipeline efficiency via genetic programming. We use TPOT builds models for extracted MobileNet network features from COVID-19 image data. The f1-score of 0.79 classifies Normal, Viral Pneumonia, and Lung Opacity.

1 Introduction

Coronaviruses are the name given to diseases caused by SARS-CoV-2 (COVID-19). The majority of those who catch the virus may experience mild to severe respiratory symptoms, but will recover without special treatment. However, some will become gravely ill and require medical care. The elderly and those with previous conditions such as cardiovascular disease, diabetes, and chronic lung disease are

F. N. Bernardo Gois · J. A. Lobo Marques (✉)
Laboratory of Applied Neurosciences, University of Saint Joseph, Estrada Marginal da Ilha Verde, 14-17, Macao SAR, China
e-mail: alexandre.lobo@usj.edu.mo

F. N. Bernardo Gois
e-mail: nauber.gois@usj.edu.mo

S. J. Fong
Faculty of Science and Technology, University of Macau, Macau SAR, China
e-mail: ccfong@um.edu.mo

J. A. Lobo Marques and S. J. Fong (eds.), *Computerized Systems for Diagnosis and Treatment of COVID-19*, https://doi.org/10.1007/978-3-031-30788-1_8

more susceptible to serious illness. COVID-19 can cause severe illness and possibly death in certain individuals. Additionally, pneumonia can refer to a lung illness caused by bacteria, viruses, or fungi. The infection causes inflammation of lung air sacs (alveoli). When fluid or pus builds in the alveoli, it is difficult to breathe. As a result, recognizing COVID-19 and Pneumonia has become an expensive and time-consuming endeavor, particularly for common individuals and radiologists. This chapter describes an AUTO-ML strategy to detect COVID on chest X-rays utilizing Transfer Learning feature extraction and the AutoML TPOT framework in order to identify lung illnesses (such as COVID or pneumonia).

Transfer learning refers to the process of adapting a model trained on one problem to a different but related problem. Transfer learning is a deep learning technique that involves training a neural network model on a problem with similar features to the target problem. The trained model is then used to inform the training of a new model on the topic of interest by reusing one or more of its layers. Transfer learning allows us to reduce the time required to train a neural network model and possibly increase its generalization accuracy. The weights in the recycled layers can serve as a baseline for the training technique, which can then be modified in light of the fresh situation. Transfer learning is understood in this context as a mechanism for initializing weights. This can be useful if the first related problem contains significantly more labeled data than the problem of interest and if the structure of both problems is similar enough to be beneficial in both cases.

Convolutional neural networks, commonly referred to as ConvNets, are a form of ANN widely employed in the field of deep learning for image analysis. CNN's are also known as Shift Invariant or Space Invariant Artificial Neural Networks because to the shared-weight architecture of the convolution kernels or filters that produce translation-equivariant outputs known as feature maps (SIANN). Contrary to predictions, the downsampling approach utilized by the vast majority of convolutional neural networks makes the input translation-sensitive. In the case of convolutional neural networks (CNNs), feature extraction is typically a result of the training phase as opposed to a separate, active step. When a convolutional neural network (CNN) is trained, it discovers how to translate input space to a latent space that may be used for classification in the final layer. This procedure of mapping is known as feature extraction. In other words, modified convolutional filters (typically by back-propagation of the error) are used to teach the hidden layers specific characteristics. Consequently, the training phase preceding the final layer is frequently known as "feature extraction." Therefore, it lacks the last-layer-only training characteristic of transfer learning.

Compressing and speeding up deep convolutional neural networks to minimize parameters, computational cost, and battery consumption is a viable method for bringing the model of these classifiers to MIoT (Medical Internet of Things) devices and simplifying its adoption. Parameters in a deep convolutional neural network were shown to be very redundant by Denil et al. [12], with minimal effect on classification performance. To estimate the information parameters of deep CNNs via singular value decompositions, Denton et al. [13] located a suitable low-rank matrix. High computational cost and additional retraining are necessary for convergence.

To further decrease the compression rate, Han et al. [19] used parameter pruning to remove unnecessary links from the pretrained network, retrained and quantized the remaining parameters, and then encoded the quantized parameters using Hoffman coding. The method is effective, but it calls for the superparameters to be tweaked by hand. For weight sharing, Chen et al. [4] employed a cheap Hash function to arrange the weights across the two adjacent layers into a Hash bucket, which minimizes the need to store information about new places and allows for the simultaneous usage of all relevant parameters. Useful information was condensed from the network model via knowledge distillation, as done by Hinton et al. [25]. After moving the relevant data to a more compact and straightforward network, both the basic and complex networks perform at around the same level [66].

In addition, several studies in this area have resulted in more efficient network compression models. Examples of such models are the fire module-based SqueezeNet [29], the depth-based separable filters of MobileNets [26], and the residual-based structure-based improvement of ShuffleNet [72], which introduces group pointwise convolution and a channel-shuffle operation [66]. MobileNet is a lightweight network that uses depthwise separable convolution to deepen the network while decreasing parameters and computation, making it more comparable to the smaller and faster VGG-16 network. MobileNet loses only 1% of its classification accuracy on the ImageNet dataset, though. However, the parameters and computational complexity of the MobileNet model must be further decreased before it can be effectively applied to mobile devices with limited memory. As a result, in MobileNet's network layer, dense blocks serve as the fundamental building component. Limiting the model's rate of growth helps keep it simple and cheap to run. Dense-MobileNets are one of the newer models that can perform well in terms of classification accuracy [66].

Many difficult AI tasks have been solved in recent years thanks to the widespread adoption of deep learning in domains like image classification [22, 37], object recognition [50], and language modeling [18]. In particular, more complicated and deep neural networks have been proposed ever since AlexNet [37] beat out all other traditional manual approaches in the 2012 ImageNet Large Scale Visual Recognition Challenge (ILSVRC) [52]. For instance, to process a 224 $\times$ 224 pixel image in VGG-16 [57] takes 15.3 billion floating-point operations, consumes approximately 500 MB of memory, and uses more than 130 million parameters. However, it is important to note that all of these models were manually created by professionals using a trial-and-error method, implying that even specialists require significant resources and effort to create models that perform effectively [24].

In order to lessen these hefty development expenses, a revolutionary concept known as automated machine learning (AML) has arisen (AutoML). The term "AutoML" can mean a number of different things depending on context. According to [73], for instance, AutoML is meant to enable domain experts to automatically construct ML applications with no need for statistical and ML skills, hence reducing the necessity for data scientists. "AutoML" is defined as "the integration of robotic process automation and machine learning" in [71]. AutoML, in a nutshell, is the process of automatically building an ML pipeline inside a constrained computing framework. AutoML has risen to prominence in both the business world and the academic community thanks to the ever-increasing capabilities of modern computers. An end-to-end ML pipeline system that is simple to use can be built with the help of an AutoML system by dynamically combining several methodologies.

COVID-19 pneumonia may be difficult to distinguish from other lung infections, especially viral or atypical ones. Interstitial lung involvement is a common symptom of viral pneumonia [16]. Influenza normally presents consolidated GGOs, bilateral consolidation, and tiny nodules [40, 44]. In contrast to seasonal influenza, the H1N1 and H5N1 strains typically manifested as quickly progressing pneumonia that ultimately resulted in ARDS—Acute Respiratory Distress Syndrome. Multifocal consolidations and diffuse regions of GGO are visible on CT scans, along with other specific structures [1, 48]. Other virus is the RSV—Respiratory Syncytial Virus (RSV), which mostly affects children and elderly patients present different patterns such as consolidations, tree-based opacity, and bronchial wall thickening [42, 47]. CT-Scans of the lungs of patients diagnosed with pneumonia originated from with herpes virus (HSV/VZV), and cytomegalovirus (CMV) usually present and asymmetric Ground-Glass opacity, associated with consolidation [6]. The automatic classification of CMV pneumonia and Pneumocystis Jirovecii pneumonia (PJP) is a hard task, since they present similar symptoms during the early stages. Differentiating and diagnosing these lung infections is crucial, since they are the most common infectious consequences in patients diagnosed with AIDS. In addition, PJP may be a significant risk of infection for patients with diabetes or under corticosteroid therapy. The most common finding in these patients are bilateral GGO. CMV presents micronodules and can be seen in either in the center or the lobulus, or even with random patterns, while PJP micronodules are more likely to be found in an apical-perihilar distribution and to have a mosaic pattern with a crazy-paving appearance [64]. Other approaches and techniques for the classification of pulmonary diseases based on the features extracted from medical imaging can be found in [9, 38, 42].

The main objective of this Chapter is to detect COVID on chest X-rays utilizing Transfer Learning feature extraction and the AutoML TPOT framework, which is a Python Optimised ML Pipeline that uses genetic programming to optimize the performance of ML pipelines, used to automatically generate models for extracted MobileNet network features.

2 Technical Background

In this Section, an overview of different Deep Learning models is presented, including:

- Transfer Learning Model—MobileNet
- Automated Machine Learning (AutoML)
- Neural Architecture Search (NAS)
- Bayesian Optimization
- Tree-Based Pipeline Optimization Tool (TPOT).

2.1 MobileNet

The latest version of Google's MobileNet network is briefly discussed here. The complexity cost and model size of the network are much reduced in MobileNetV1 thanks to Depthwise Separable Convolution, making it well-suited to Mobile devices or other devices with limited processing capability.

MobileNetV1 consists of two distinct layers. Lightweight filtering is carried out by the first layer, a depthwise convolution, using a single convolutional filter for each input channel. To generate novel features, the second layer performs an 11 convolution, also known as a pointwise convolution, by linear computing combinations of the input channels.

In an effort to optimize CNN performance on mobile devices, MobileNetV2 was developed. It uses a novel structure compared to traditional CNNs, with interconnections between the bottleneck layers. The intermediate expansion layer employs deep folds to remove nonlinear characteristics as an additional bonus. 19 bottleneck layers follow the 32 initial convolution layers in MobileNetV2 [33].

MobileNetV2 adds a new, improved module that has an inverted residual structure. The previous non-linearities in thin layers have been eliminated. State-of-the-art performances are reached for both object detection and semantic segmentation using MobileNetV2 as the foundation for feature extraction.

To decrease processing in the first few layers, MobileNets are constructed largely from depthwise separable convolutions, which were first described in [55] and later employed in Inception models [30]. The potential of highly factorized networks was demonstrated by flattened networks [32], which construct a network from fully factorized convolutions. Similar factorized convolution and topological connections are introduced in a publication called Factorized Networks [65], which is unrelated to the current work. In subsequent work, the Xception network [5] showed how to scale up depthwise separable filters to outperform Inception V3 networks. Squeezenet [29], which employs a bottleneck strategy in its network construction, is another example of a tiny network. The networks of reduced computation include not only deep-fried convnets [70] but also structured transform networks [58]. Shrinkage, factorization, or compression of pre-trained networks is an alternative method for achieving tiny networks. Product quantization [68], hashing [4], pruning, vector quantization, and Huffman coding [19] are only a few of the compression techniques presented in the literature. In addition, several factorizations have been proposed to quicken pre-trained networks [31, 39]. Distillation [25] is another technique for training tiny networks, and it does it by utilizing a bigger network to instruct a smaller one [26].

Kaya and Gürsoy use five distinct pre-trained CNN architectures (VGG16, DenseNet, MobileNetV2, InceptionV3, and ResNet) were evaluated to determine which one performed best. Results reveal that MobileNetV2 is more effective after training for 100 epochs with 80% of the photos. With this in mind, we suggest using a deep transfer learning method based on MobileNetV2 with innovative fine-tuning techniques to identify COVID-19 disorders in chest X-ray pictures. To assess the proposed model, the study assembled a sizable dataset by merging two existing public resources: the COVID-19 Radiography Database [34] and the Chest X-ray Images (Pneumonia) [8]. Normal, pneumonia and COVID-19 were the three categories included in the dataset [33].

Rab Ratul et al. [49] present RAM-Net, a DNN that combines MobileNet with dilated depthwise separable convolution (DDSC), residual blocks, and attention-augmented convolution. The network was trained and tested with images from the COVIDx dataset, which is among the most widely used public CXR image datasets. By training the network on new, potentially worrisome chest X-ray images, the study was able to accurately identify cases of COVID-19 viral infection using this model. To the best of our knowledge, our network achieved the highest accuracy to date on the COVIDx test dataset, with a 95.33% overall accuracy, a sensitivity of 92%, and a precision of 99% for COVID-19 cases. Finally, the research conducted an audit of RAM-Net based on the Grad-interpretation CAMs to show that our suggested architecture detects SARS-CoV2 (COVID-19) viral infection without relying on non-essential information [49].

2.2 *Automated Machine Learning—AutoML*

The demand for data-driven software is on the rise, and machine learning (ML) is a popular approach considered to achieve efficient data analysis. The rapid adoption of ML technology, together with exponential growth in data generation and a dearth of data science talent have placed an increased burden on the industry to keep up with rapidly evolving design lifecycles. Due to the first-order design aim, research initiatives are addressing this challenge using the concept of multiple pipelines of machine learning algorithms, which was called AutoML - Automated Machine Learning [7, 10, 14, 15, 20, 43, 51, 53, 54, 59–63, 69].

Data preprocessing, feature engineering, model building, and analysis are only some of the steps in the AutoML pipeline. Model creation strategies may be further subcategorized into optimization and search strategies. Different ML techniques, have different search spaces strategies and are commonly divided into two groups: the traditional Machine-Learning models, including kNN (k-Nearest Neighbors), SVM (Support Vector Machines), and others, and models based on artificial neural networks [23].

In the end, this becomes an optimization task and several approaches may be considered. Researchers have handled this optimization problem using several different methods. One of the approaches is based on Bayesian Optimization [35, 36], which is focused on developing probabilistic models to evaluate multiple hyperparameter configurations and their corresponding performance. In addition, AutoML approaches may also consider Genetic Algorithms to perform the optimization task, such as TPOT.

A popular Python library considering the Bayesian approach is the "auto-sklearn", which uses a sequential model-based optimization technique based on random forests to optimize the overall configuration of its algorithms. It uses meta-learning and follows a previously optimized dataset that is most similar to the supplied dataset and then bootstraps the iterative optimization process with the configuration of the well-known dataset [20]. Other libraries are also available and achieve significant results in the literature, according to the application it was designed for, such as H2O.ai [41], TPOT (which is the focus of the current chapter and uses genetic algorithms for the optimization process), Auto-Keras (from the popular Keras framework), and AutoML framework from Google. The last two consider the NASP—Neural Architecture Search Problem approach, which is presented as follows.

2.3 Neural Architecture Search Problem

Over the past few years, Deep Learning has enabled remarkable progress in a wide range of applications, including image and speech processing, and translation tasks. Traditionally, architects have worked on designs by hand, a process that may be slow and fraught with human error. As a result, automated neural search techniques are garnering more and more attention [20, 67].

Automation of machine learning is a new possibility for optimizing these multiple existing approaches from the concept of automating architectural engineering. This is what Neural Architecture Search (NAS) does. The Neural Architecture Search algorithm will routinely look for the best possible neural network architecture and its associated tuning parameters. On several tasks, such as image classification, object detection, and semantic segmentation, NAS approaches have already surpassed those of manually constructed architectures [3]. NAS is a subfield of AutoML (Hutter et al. 2019) and has significant overlap with hyperparameter optimization (Feurer and Hutter 2019) and meta-learning [27]. We categorize NAS's approaches based on three dimensions: research space, research technique, and performance estimation strategy [20, 67].

For a better understanding, a formal definition of the model is provided. Assuming that our neural network solution belongs to the search space SNN, the input data D is then split into two blocks: $D_{training}$ and $D_{validation}$, with a proper cost function C. The optimal solution for the ANN model $snn \in SNN$, which could achieve the lowest C for the dataset D can be stated as follows [20]:

$$f^* = arg(min(Cost\left(snn\left(\theta^*\right), D_{val}\right))),\ snn \in SNN \quad (1)$$

$$\theta^* = arg(min(L\left(f\left(\theta\right), D_{train}\right))),\ \theta \quad (2)$$

Actually, the concept of Cost will be presented as a model performance measure, such as MSE (Mean Squared Error). If other metrics, such as Accuracy or F1-score are considered, so the optimization problem becomes a Maximization problem and the max function shall be considered in the equations. The search space SNN will consider all virtually possible neural networks, which are based on initial models able to achieve a solution to the problem under study.

2.4 Bayesian Optimization

Using the Bayes Theorem is a popular approach to solving an optimization problem. It is a foundational probabilistic model of finding an objective function to achieve a minimum or maximum, according to the desired optimal solution. It consists of testing multiple functions, searching for the final optimal result, which is usually called objective function.

This searching strategy makes this optimization approach suitable for complex problems hard to be modeled, such as Machine Learning algorithms. It consists in the execution of three main phases. The first one is the update phase, consisting on training a general Gaussian model. The second phase is called generation and defines a new model to optimize the solution. Finally, the third phase is the observation, which evaluates the performance of the model created in the previous phase.

One example of a general algorithm for Bayesian Optimization is presented as follows.

Listing 1 Algorithm 1 Bayesian optimization

```
Input: Objective function f()
Input: Acquisition function a()
Initialize the Gaussian process G
for i = 1, 2, . . . do
Sample point: xt <- argmax a(G(x))
Evaluate new point: yt <- f(xt)
Update the Gaussian process: G<- G|(xt, yt)
end for
```

2.5 TPOT AutoML

The study of how machines can acquire new skills without being given new instructions is known as machine learning [56]. Despite this common belief, those who work in the field of machine learning know that building effective machine learning pipelines typically requires technical expertise, including understanding the algorithms and models, effective comprehension of the problem under analysis, and some level of manual comparison to determine the best approach [45]. Machine learning, contrary to common opinion, still relies heavily on explicit programming. In order to solve this problem, numerous machine learning automation techniques have been created [28].

The machine learning pipeline considered in this Chapter is the Tree-based Pipeline Optimization Tool (TPOT) [45]. TPOT is an automatic build and tune ML pipeline based on Genetic programming (GP), which is an automatic coding development toolusing evolutionary computation approach.

TPOT uses GP to optimize machine learning pipelines [2]. The automatic construction of compact pipelines that exceed fundamental machine learning studies with high-accuracy, was demonstrated to be possible by merging GP with Pareto optimization. From clinical analysis to image classification, the literature demonstrates a benchmark to more than 150 supervised classification tasks [45].

The scikit-learn machine learning library for Python brings a TPOT container [46]. This means that each TPOT's machine learning pipeline represents a distinct machine learning technique, such as a feature extraction or a supervised classifier. Except for XGBoost, several machine learning methods are implemented in this library. Detailed information on these algorithms and how they are used in TPOT can be found in the official scikit-learn documentation [46] or the work of Hastie et al.[21]. Operators for Supervised Classification, Pre-processing Features and Selecting Features. In addition, TPOT provide a data-set-combination operator, which enables many updated versions of the data set to be combined into a single data set.

The GP primitives are the basic operators for the TPOT container. Then it uses trees to incorporate them into a machine learning pipeline. A tree-based pipeline receives two copies of a data set, which are then iterative updated by each operator before being combined into a single data set for use in classifications. Machine learning pipelines of any conceivable structure are possible, provided all operators take a data set as input and return the updated data set as output. Because of this, GP trees are a highly adaptable framework for modeling ML workflows.

TPOT employ the GP algorithm implemented in the Python module DEAP [17] to automatically create and optimize these tree-based pipelines. TPOT is a general-purpose GP algorithm that uses the usual GP steps: The GP approach starts by producing 100 randomly selected pipelines and then evaluating their performance through cross-validation accuracy. The NSGA-II selection technique [11] was proposed to consider the best twenty pipelines for each generation. These pipelines are chosen so that they optimize classification accuracy on the data set while limiting the number of operators in the pipeline. The GP model for each of the top 20 selected pipelines has five offspring in the next generation, of which 5% undergo one-point crossover with another offspring and 90% undergo through random point mutations (insert or shrink) with a 33% chance each [11]. At the end of 100 generations, the algorithm chooses the highest-accuracy pipeline from a Pareto ranking as the representative for the "best" pipeline from the run. Then it repeats the evaluate-select-crossover-mutate process to add and tune the solutions which are able to improve the model accuracy while it automatically removes solutions with poor classification performance.

Models, feature pre-possessing/selection strategies, and other combinations can be put to the test with the help of the TPOT package. The employed combinations are referred to as a TPOT pipeline. This one-of-a-kind item contains all the data necessary to recreate the model under test. The user can also choose which models and techniques will be employed during the training process. As a result, the user can choose from a wider variety of methods while training the model. Similar feature selection and data reprocessing methods were employed in this investigation as were in the prior one.

3 Methodology

The proposed procedure includes five steps (Fig. 1). The initial step is to choose the proper database. The base was chosen based on the classification possibilities between Normal, COVID, and other respiratory diseases. Using the pre-trained MobileNet network, the second and third processes consist of producing an array containing the desired features. Editing the final layer of the network and adding a layer of 120 neurons with the sigmoidal activation function yielded the array.

The proposed method use the MobileNet V2 base model is used to extract a collection of useful features from COVID image dataset. Listing 2 present the code of the neural network to extract features from COVID image dataset. The last layer have 102 nodes.

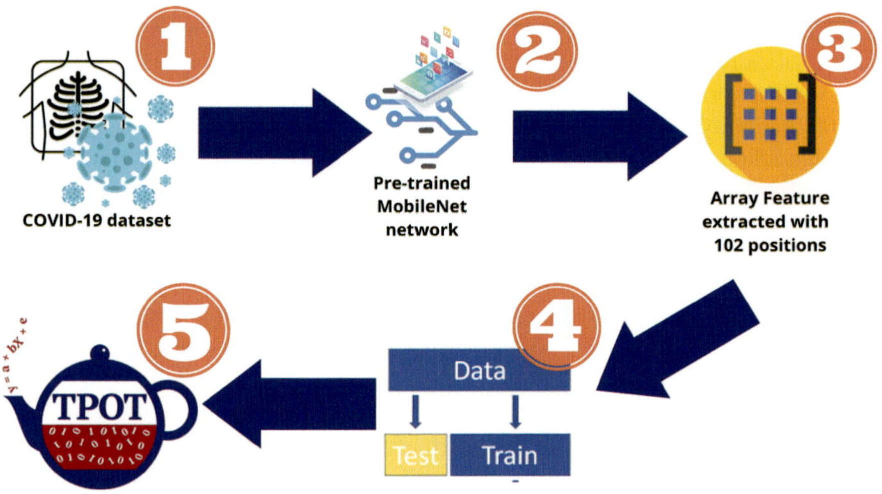

Fig. 1 Proposed method

```
import tensorflow as tf
from tensorflow import keras

IMG_SHAPE = (70, 70, 3)

base_model = tf.keras.applications.MobileNet(input_shape=IMG_SHAPE
    , include_top=False)
base_model.trainable = False

base_model = tf.keras.Sequential([
  base_model,
  keras.layers.GlobalAveragePooling2D(),
  keras.layers.Dense(102, activation='sigmoid')
])

base_model.summary()
```

Listing 2 Feature Extraction with MobileNet pre-trained network

Listing 3 present the code to extract features from COVID image dataset.

```
x_train_features=base_model.predict(x_train)
x_test_features=base_model.predict(x_test)
```

Listing 3 Obtain features with pretraind MobileNet model

TPOT analyzes a wide variety of preprocessors, feature constructors and selectors, models, and hyperparameters to find those that reduce the model predictions error. For our approach, it examines the subsequent:

- linear models (like Logistic Regression)
- naive bayes models (like Bernoulli NB, Gaussian NB, MultinomialNB)
- tree models (like Decision Tree Classifier)
- ensemble models (like Random Forest Classifier)
- SVM models (like LinearSVC) XGBoost models.

Listing 4 present the code to train TPOT model.

```
tpot = TPOTClassifier(generations=2, population_size=15, verbosity=2)
tpot.fit(x_train_features, y_train)
```

Listing 4 Using TPOT to classify extracted features

3.1 Database

A team of researchers from Qatar University, Doha, Qatar, and the University of Dhaka, Bangladesh, along with their collaborators from Pakistan and Malaysia and medical physicians, established a database of COVID-19-positive chest X-Rays (CXR), together with patients diagnosed with viral pneumonia (non-COVID), and normal CXR images [8] (Fig. 2).

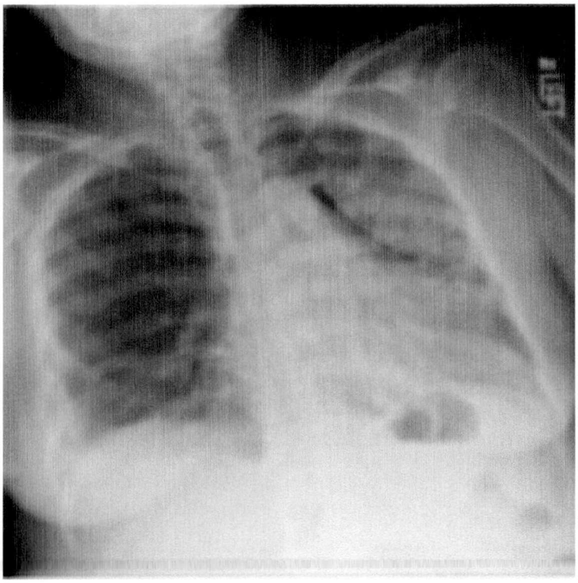

Fig. 2 Example of lung opacity

3.2 *Evaluation Criteria*

This chapter will consider the confusion matrix and other three performance indicators to evaluate the proposed classification systems: Precision, Recall, and Accuracy.

3.2.1 Confusion Matrix

First, the Confusion Matrix is a common representation for the evaluation of false negatives and false positives identified by the proposed classifier. In a more formal way, the matrix can be stated as follows.

Let MI(x,y) : $\mathbb{R}^2 \rightarrow \mathbb{R}$ represent a medical image and O(MI(x,y)):$\mathbb{R}^2 \rightarrow \Omega, \Omega =$ 0, 1 a classification of the medical image MI(x,y). The classification criteria are then presented considering the classification goal as G, and the result achieved by the system as R Popovic2007:

- true positive: $G(x, y) = 1 \wedge R(x, y) = 1$,
- false positive: $G(x, y) = 0 \wedge R(x, y) = 1$,
- true negative: $G(x, y) = 0 \wedge R(x, y) = 0$,
- false negative: $G(x, y) = 1 \wedge R(x, y) = 0$,

3.2.2 Precision

The Precision is a metric to evaluate the system performance from the perception of the number of false positives, compared to the number of true positive cases and is given by:

$$P = \frac{TP}{TP + FP}$$

3.2.3 Recall

The Recall (R) is a metric to evaluate the system performance from the perception of the number of false negatives (FN), compared to the number of true positive (TP) cases, i.e. it measures the system performance not to miss one TP case and is given by:

$$R = \frac{TP}{TP + FN}$$

3.2.4 Accuracy

Accuracy is a popular metric for evaluating a model's performance, considering both positive and negative classes and considering true and false positives (TP and FP) together with true and false negatives (TN and FN) in the same metric. The equation is presented as follows:

$$Accuracy = \frac{TP + TN}{TP + TN + FP + FN}$$

4 Results

Table 1 presents the results of the experiments with the TPOT approach. The classes are identified as [0, 1, 2], where 0 indicates healthy patients, 1 indicated Non-COVID-19 pulmonary disease, and 2 indicates COVID-19 diagnosis. The metrics Precision, Recall, and F1-scores are presented.

The maximum Precision was 0.80, and it was achieved for class 0, which represents healthy patients. For patients with COVID-19, the Precision was 0.79 (class 2), while for other pulmonary diseases (class 1) was 0.78.

Considering the Recall, the best result was 0.87, and it was achieved again for class 0, which represents healthy patients. For patients of class 2 (with COVID-19), the Recall was 0.75, while for other pulmonary diseases (class 1) was 0.70.

Table 1 TPOT experiment results

class	Precision	Recall	f1-score	Support
0	0.80	0.87	0.83	905
1	0.78	0.70	0.74	492
2	0.79	0.75	0.77	603
Accuracy			0.79	2000
Macro avg	0.79	0.77	0.78	2000
Weighted avg	0.79	0.79	0.79	2000

The same behavior was presented for the F1-score metric. The best F1-score was 0.83 for class 0 (healthy patients). For patients with COVID-19 (class 2), the precision was 0.74, while for other pulmonary diseases (class 1) was 0.74.

The overall Accuracy of the system was 79%, which may indicate that the TPOT AutoML model may be a suitable approach for classifying multi-class Chest X-Ray.

5 Conclusion

SARS-CoV-2 causes coronaviruses (COVID-19). Most patients may be able to recover without therapeutic interventions, despite mild to severe respiratory symptoms, while others may evolve to severe states and need medical intervention. Some groups are more susceptible to severe sickness, including the elderly and those with cardiovascular, diabetes, and chronic lung conditions. COVID-19 can kill certain people. Bacteria, viruses, and fungi can also cause pneumonia. Infection inflames lung air sacs (alveoli). Alveoli fluid or pus makes breathing harder. Thus, diagnosing COVID-19 and pneumonia is costly and time-consuming, especially for radiologists and the general public. This chapter describes an AUTO-ML technique to detect COVID on chest X-rays using Transfer Learning feature extraction and the AutoML TPOT framework to identify lung diseases (such as COVID or pneumonia). Transfer learning adapts a model learned from one problem to a related problem. Transfer learning is a deep learning method that trains a neural network model on a similar task. Reusing one or more layers of the learned model helps train a new model on the issue. Transfer learning shortens neural network model training and may improve generalization accuracy. The recycled layers' weights can be used as a baseline for training, which can be adjusted for the new situation. Transfer learning initializes weights. If the first linked problem has more labeled data than the problem of interest and if their structures are comparable enough to benefit both problems, this can be advantageous. Deep learning uses ConvNets to analyze images. The shared-weight architecture of convolution kernels or filters produces translation-equivariant feature maps, making CNNs Shift Invariant or Space Invariant Artificial Neural Networks (SIANN). Contrary to assumptions, most convolutional neural networks use down-sampling, which makes input translation-sensitive. Convolutional neural networks

(CNNs) extract features during training. Convolutional neural networks (CNNs) learn how to transform input space to latent space for classification in the final layer. Feature extraction maps. Thus, updated convolutional filters (usually by error backpropagation) teach hidden layers certain traits. "Feature extraction" is the training phase before the final layer. It lacks transfer learning's last-layer-only training. Deep convolutional neural networks can be compressed and accelerated to reduce parameters, computation cost, and energy consumption for Medical IoT applications. Denil et al. found that deep convolutional neural network parameters were redundant and did not generate any impact on the classification performance. Denton et al. used singular value decompositions to determine a technique to estimate deep CNN parameters. Convergence requires expensive computation and retraining. Han et al. [19] pruned superfluous links from pre-trained networks, retraining and proposing a quantization of the network parameters, and, finally, using Hoffman coding to reduce compression of the quantized parameter. It works, however, hyperparameters must be adjusted manually. Several studies have improved network compression models. SqueezeNet, MobileNets, and ShuffleNet, which introduces group pointwise convolution and a channel-shuffle operation, are examples of such models. MobileNet, a lightweight network, uses depthwise separable convolution to deepen the network while reducing parameters and processing. MobileNet loses only 1% classification accuracy on ImageNet. The MobileNet model can be used on mobile devices with limited memory, but its parameters and computational complexity must be reduced. Thus, dense blocks are MobileNet's network layer's building blocks. Limiting model growth simplifies and lowers costs. Dense-MobileNets, a new classification model, performs well. Automated machine learning (AML) reduces these high development costs (AutoML). "AutoML" has many meanings depending on context. According to [73], AutoML allows domain experts to automatically build ML applications without statistical or ML capabilities, minimizing the requirement for data scientists. "AutoML" is "the merging of robotic process automation and machine learning" in [71]. AutoML builds an ML pipeline in a limited computing architecture. Modern computers have made AutoML popular in industry and academia. AutoML may dynamically combine different approaches to create a simple end-to-end ML pipeline solution. Python Optimised ML Pipeline (TPOT) is a Python-based ML tool that optimizes pipeline efficiency via genetic programming. TPOT builds models for extracted MobileNet network features. The experiments obtain a f1-score of 0.79 to classifies Normal, Viral Pneumonia, and Lung Opacity.

References

1. Ajlan AM, Quiney B, Nicolaou S, Muller NL (2009) Swine-origin influenza a (h1n1) viral infection: radiographic and CT findings. Am J Roentgenol 193(6):1494–1499
2. Banzhaf W, Nordin P, Keller RE, Francone FD (1998) Genetic programming: an introduction: on the automatic evolution of computer programs and its applications. Morgan Kaufmann Publishers Inc

3. Bender G, Kindermans PJ, Zoph B, Vasudevan V, Le Q (2018) Understanding and simplifying one-shot architecture search. In: The 35th international conference on machine learning ICML 2018, vol 2, pp 883–893
4. Chen W, Wilson J, Tyree S, Weinberger K, Chen Y (2015) Compressing neural networks with the hashing trick. In: International conference on machine learning, PMLR, pp 2285–2294
5. François Chollet. Xception: Deep learning with depthwise separable convolutions. In *Proceedings of the IEEE conference on computer vision and pattern recognition*, pages 1251–1258, 2017
6. Chong S, Kim TS, Cho EY (2010) Herpes simplex virus pneumonia: high-resolution ct findings. Br J Radiol 83(991):585–589
7. Chouhan V, Singh SK, Khamparia A, Gupta D, Tiwari P, Moreira C, Damaševičius R, de Albuquerque VHC (2020) A novel transfer learning based approach for pneumonia detection in chest X-ray images. Appl Sci (Switzerland) 10(2)
8. Chowdhury MEH, Rahman T, Khandakar A, Mazhar R, Kadir MA, Mahbub ZB, Islam KR, Khan KS, Iqbal A, Al Emadi N et al (2020) Can AI help in screening viral and covid-19 pneumonia? IEEE Access 8:132665–132676
9. Cozzi D, Cavigli E, Moroni C, Smorchkova O, Zantonelli G, Pradella S, Miele V (2021) Ground-glass opacity (GGO): a review of the differential diagnosis in the era of covid-19. Jpn J Radiol 39(8):721–732
10. De Souza RWR, De Oliveira JVC, Passos LA, Ding W, Papa JP, Albuquerque V (2019) A novel approach for optimum-path forest classification using fuzzy logic. IEEE Trans Fuzzy Syst 6706(c):1–1
11. Deb K, Pratap A, Agarwal S, Meyarivan TAMT (2002) A fast and elitist multiobjective genetic algorithm: Nsga-ii. IEEE Trans Evol Comput 6(2):182–197
12. Denil M, Shakibi B, Dinh L, Ranzato M, De Freitas N (2013) Predicting parameters in deep learning. Advances in neural information processing systems, 26
13. Denton EL, Zaremba W, Bruna J, LeCun Y, Fergus R (2014) Exploiting linear structure within convolutional networks for efficient evaluation. Advances in neural information processing systems, 27
14. Ding W, Abdel-Basset M, Eldrandaly KA, Abdel-Fatah L, de Albuquerque VHC (2020) Smart Supervision of cardiomyopathy based on fuzzy harris hawks optimizer and wearable sensing data optimization: a new model. IEEE Trans Cybernet, pp 1–15
15. Dourado CMJM, Da Silva SPP, Da Nobrega RVM, Reboucas Filho PP, Muhammad K, De Albuquerque VHC (2020) An open IoHT-based deep learning framework for online medical image recognition. IEEE J Sel Areas Commun XX(X):1–8
16. Duzgun SA, Durhan G, Demirkazik FB, Akpinar MG, Ariyurek OM (2020) Covid-19 pneumonia: the great radiological mimicker. Insights into Imaging 11(1):1–15
17. Fortin F-A, De Rainville F-M, Gardner Gardner M-A, Parizeau M, Gagné C (2012) Deap: evolutionary algorithms made easy. J Mach Learn Res 13(1):2171–2175
18. Gong C, He D, Tan X, Qin T, Wang L, Liu T-Y (2018) Frage: frequency-agnostic word representation. Advances in neural information processing systems, 31
19. Han S, Mao H, Dally WJ (2015) Deep compression: compressing deep neural networks with pruning, trained quantization and huffman coding. arXiv:1510.00149
20. Han T, Bernardo Gois FN, Oliveira R, Prates LR, de Almeida Porto MM (2021) Modeling the progression of covid-19 deaths using kalman filter and automl. Soft Computing, pp 1–16
21. Hastie T, Tibshirani R, Friedman JH, Friedman JH (2009) The elements of statistical learning: data mining, inference, and prediction, vol 2. Springer
22. He K, Zhang X, Ren S, Sun J (2016) Deep residual learning for image recognition. In: Proceedings of the IEEE conference on computer vision and pattern recognition, pp 770–778
23. He X, Zhao K, Chu X (2019) AutoML: a survey of the state-of-the-art. (Dl)
24. He X, Zhao K, Chu X (2021) Automl: a survey of the state-of-the-art. Knowl-Based Syst 212:106622
25. Hinton G, Vinyals O, Dean J et al (2015) Distilling the knowledge in a neural network 2(7). arXiv:1503.02531

26. Howard AG, Zhu M, Chen B, Kalenichenko D, Wang W, Weyand T, Andreetto M, Adam H (2017) Mobilenets: efficient convolutional neural networks for mobile vision applications. arXiv:1704.04861
27. Hutter F, Kotthoff L, Vanschoren J (2019) Automated machine learning: methods, systems, challenges. Springer Nature
28. Hutter F, Lücke J, Schmidt-Thieme L (2015) Beyond manual tuning of hyperparameters. KI-Künstliche Intelligenz 29(4):329–337
29. Iandola FN, Han S, Moskewicz MW, Ashraf K, Dally WJ, Keutzer K (2016) Squeezenet: Alexnet-level accuracy with 50x fewer parameters and < 0.5 mb model size. arXiv:1602.07360
30. Ioffe S, Szegedy C (2015) Batch normalization: Accelerating deep network training by reducing internal covariate shift. In: International conference on machine learning. PMLR, pp 448–456
31. Jaderberg M, Vedaldi A, Zisserman A (2014) Speeding up convolutional neural networks with low rank expansions. arXiv:1405.3866
32. Jin J, Dundar A, Culurciello E (2014) Flattened convolutional neural networks for feedforward acceleration. arXiv:1412.5474
33. Kaya Y, Gürsoy E (2023) A mobilenet-based CNN model with a novel fine-tuning mechanism for covid-19 infection detection. Soft Computing, pp 1–15
34. Kermany DS, Goldbaum M, Cai W, Valentim CCS, Liang H, Baxter SL, McKeown A, Yang G, Wu X, Yan F et al (2018) Identifying medical diagnoses and treatable diseases by image-based deep learning. Cell 172(5):1122–1131
35. Komer B, Bergstra J, Eliasmith C (2014) Hyperopt-sklearn: automatic hyperparameter configuration for scikit-learn. In: Proceedings of the 13th python in science conference, (Scipy), pp 32–37
36. Kotthoff L, Thornton C, Hoos HH, Hutter F, Leyton-Brown K (2017) Auto-weka 2.0: automatic model selection and hyperparameter optimization in weka. J Mach Learn Res 18(1):826–830
37. Krizhevsky A, Sutskever I, Hinton GE (2017) Imagenet classification with deep convolutional neural networks. Commun ACM 60(6):84–90
38. Lansbury L, Lim B, Baskaran V, Lim WS (2020) Co-infections in people with covid-19: a systematic review and meta-analysis. J Infect 81(2):266–275
39. Lebedev V, Ganin Y, Rakhuba M, Oseledets I, Lempitsky V (2014) Speeding-up convolutional neural networks using fine-tuned CP-decomposition. arXiv:1412.6553
40. Leung AN, Gosselin MV, Napper CH, Braun SG, Hu WW, Wong RM, Gasman J (1999) Pulmonary infections after bone marrow transplantation: clinical and radiographic findings. Radiology 210(3):699–710
41. Lobo Marques JA, Nauber Bernardo Gois F, Xavier-Neto J, James Fong S (2021) Artificial intelligence prediction for the COVID-19 data based on LSTM neural networks and H2O autoML. In: Lobo Marques JA, Nauber Bernardo Gois F, Xavier-Neto J, James Fong S (eds) Predictive models for decision support in the COVID-19 crisis, Springer briefs in applied sciences and technology. Springer International Publishing, Cham, pp 69–87
42. Miller WT Jr, Mickus TJ, Barbosa E Jr, Mullin C, Van Deerlin VM, Shiley KT (2011) CT of viral lower respiratory tract infections in adults: comparison among viral organisms and between viral and bacterial infections. Am J Roentgenol 197(5):1088–1095
43. Muhammad K, Khan S, Del Ser J, de Albuquerque VHC (2020) Deep learning for multigrade brain tumor classification in smart healthcare systems: a prospective survey. IEEE Trans Neural Netw Learn Syst, pp 1–16
44. Oikonomou A, Muller NL, Nantel S (2003) Radiographic and high-resolution CT findings of influenza virus pneumonia in patients with hematologic malignancies. Am J Roentgenol 181(2):507–511
45. Olson RS, Moore JH (2016) Tpot: a tree-based pipeline optimization tool for automating machine learning. In: Workshop on automatic machine learning. PMLR, pp 66–74
46. Pedregosa F, Varoquaux G, Gramfort A, Michel V, Thirion B, Grisel O, Blondel M, Prettenhofer P, Weiss R, Dubourg V et al (2011) Scikit-learn: machine learning in python. J Mach Learn Res 12:2011

47. Pierangeli A, Scagnolari C, Antonelli G (2018) Respiratory syncytial virus. Minerva Pediatr 70(6):553–565
48. Qureshi NR, Hien TT, Farrar J, Gleeson FV (2006) The radiologic manifestations of h5n1 avian influenza. J Thoracic Imaging 21(4):259–264
49. Ratul MAR, Elahi MT, Yuan K, Lee W (2020) Ram-net: a residual attention mobilenet to detect covid-19 cases from chest x-ray images. In: 2020 19th IEEE international conference on machine learning and applications (ICMLA). IEEE, pp 195–200
50. Redmon J, Divvala S, Girshick R, Farhadi A (2016) You only look once: unified, real-time object detection. In: Proceedings of the IEEE conference on computer vision and pattern recognition, pp 779–788
51. Rodrigues MB, Da Nóbrega RVM, Alves SSA, Rebouças Filho PP, Duarte JBF, Sangaiah AK, De Albuquerque VHC (2018) Health of things algorithms for malignancy level classification of lung nodules. IEEE Access 6:18592–18601
52. Russakovsky O, Deng J, Hao S, Krause J, Satheesh S, Ma S, Huang Z, Karpathy A, Khosla A, Bernstein M et al (2015) Imagenet large scale visual recognition challenge. Int J Comput Vis 115(3):211–252
53. Santos MAG, Munoz R, Olivares R, Rebouças Filho PP, Del Ser J, de Albuquerque VHC (2020) Online heart monitoring systems on the internet of health things environments: a survey, a reference model and an outlook. Inf Fusion 53(December 2018):222–239
54. Selvachandran G, Quek SG, Lan LTH, Hoang Son L, Giang NL, Ding W, Abdel-Basset M, Albuquerque VHC (2019) A new design of mamdani complex fuzzy inference system for multi-attribute decision making problems. IEEE Trans Fuzzy Syst 6706(c):1–1
55. Sifre L, Mallat S (2014) Rigid-motion scattering for image classification [PhD thesis]. Ecole Polytechnique
56. Simon P (2013) Too big to ignore: the business case for big data, vol 72. Wiley
57. Simonyan K, Zisserman A (2014) Very deep convolutional networks for large-scale image recognition. arXiv:1409.1556
58. Sindhwani V, Sainath T, Kumar S (2015) Structured transforms for small-footprint deep learning. In: Advances in neural information processing systems, 28
59. Sodhro AH, Fortino G, Pirbhulal S, Lodro MM, Shah MA (2017) 16 energy efficiency in wireless body sensor networks. In: Networks of the future: architectures, technologies, and implementations, pp 339
60. Sodhro AH, Li Y, Shah MA (2016) Energy-efficient adaptive transmission power control for wireless body area networks. IET Commun 10(1):81–90
61. Sodhro AH, Luo Z, Sodhro GH, Muzamal M, Rodrigues JJPC, de Albuquerque VHC (2019) Artificial intelligence based QOS optimization for multimedia communication in IOV systems. Future Generat Comput Syst 95:667–680
62. Sodhro AH, Pirbhulal S, Luo Z, de Albuquerque VHC (2019) Towards an optimal resource management for IoT based green and sustainable smart cities. J Cleaner Prod 220:1167–1179
63. Sodhro AH, Pirbhulal S, Sodhro GH, Muzammal M, Zongwei L, Gurtov A, de Macêdo ARL, Wang L, Garcia NM, de Albuquerque VHC (2020) Towards 5g-enabled self adaptive green and reliable communication in intelligent transportation system. IEEE Trans Intell Transp Syst
64. Vogel MN, Brodoefel H, Hierl T, Beck R, Bethge WA, Claussen CD, Horger MS (2007) Differences and similarities of cytomegalovirus and pneumocystis pneumonia in HIV-negative immunocompromised patients-thin section ct morphology in the early phase of the disease. Br J Radiol 80(955):516–523
65. Wang M, Liu B, Foroosh H (2017) Factorized convolutional neural networks. In: Proceedings of the IEEE international conference on computer vision workshops, pp 545–553
66. Wang W, Li Y, Zou T, Wang X, You J, Luo Y (2020) A novel image classification approach via dense-mobilenet models. Mobile Inf Syst 2020
67. Wistuba M, Rawat A, Pedapati T (2019) A Surv Neural Arch Search. 20:1–21
68. Wu J, Leng C, Wang Y, Hu Q, Cheng J (2016) Quantized convolutional neural networks for mobile devices. In: Proceedings of the IEEE conference on computer vision and pattern recognition, pp 4820–4828

69. Yakovlev A, Moghadam HF, Moharrer A, Cai J, Chavoshi N, Varadarajan V, Agrawal SR, Idicula S, Karnagel T, Jinturkar S et al (2020) Oracle automl: a fast and predictive automl pipeline. In: Proc VLDB Endow 13(12):3166–3180
70. Yang Z, Moczulski M, Denil M, De Freitas N, Smola A, Song L, Wang Z (2015) Deep fried convnets. In: Proceedings of the IEEE international conference on computer vision, pp 1476–1483
71. Yao Q, Wang M, Chen Y, Dai W, Li Y-F, Tu W-W, Yang Q, Yu Y (2018) Taking human out of learning applications: a survey on automated machine learning. arXiv:1810.13306
72. Zhang X, Zhou X, Lin M, Sun J (2018) Shufflenet: an extremely efficient convolutional neural network for mobile devices. In: Proceedings of the IEEE conference on computer vision and pattern recognition, pp 6848–6856
73. Zoller M-A, Huber MF (2019) Benchmark and survey of automated machine learning frameworks. Learning

Evaluation of ECG Non-linear Features in Time-Frequency Domain for the Discrimination of COVID-19 Severity Stages

Pedro Ribeiro, Daniel Pordeus, Laíla Zacarias, Camila Leite, Manoel Alves Neto, Arnaldo Aires Peixoto Jr, Adriel de Oliveira, João Paulo Madeiro, Joao Alexandre Lobo Marques, and Pedro Miguel Rodrigues

Abstract In 2020, the World Health Organization declared the Coronavirus Disease 19 a global pandemic. While detecting COVID-19 is essential in controlling the disease, prognosis prediction is crucial in reducing disease complications and patient mortality. For that, standard protocols consider adopting medical imaging tools to analyze cases of pneumonia and complications. Nevertheless, some patients develop different symptoms and/or cannot be moved to a CT-Scan room. In other cases, the devices are not available. The adoption of ambulatory monitoring examinations, such as Electrocardiography (ECG), can be considered a viable tool to address the patient's cardiovascular condition and to act as a predictor for future disease outcomes. In this investigation, ten non-linear features (*Energy*, *Approximate Entropy*, *Logarithmic Entropy*, *Shannon Entropy*, *Hurst Exponent*, *Lyapunov Exponent*, *Higuchi Fractal Dimension*, *Katz Fractal Dimension*, *Correlation Dimension* and *Detrended Fluctuation Analysis*) extracted from 2 ECG signals (collected from 2 different patient's positions). Windows of 1 second segments in 6 ways of windowing signal analysis crops were evaluated employing statistical analysis. Three categories of outcomes are considered for the patient status: Low, Moderate, and Severe, and

P. Ribeiro · P. Miguel Rodrigues
CBQF–Centro de Biotecnologia e Química Fina–Laboratório Associado, Escola Superior de Biotecnologia, Universidade Católica Portuguesa, Rua de Diogo Botelho 1327, 4169-005 Porto, Portugal
e-mail: s-pmsbribeiro@ucp.pt

P. Miguel Rodrigues
e-mail: pmrodrigues@ucp.pt

D. Pordeus · J. Paulo Madeiro
Department of Computing, Federal University of Ceará, Fortaleza, Ceará, Brazil
e-mail: pordeus@alu.ufc.br

J. Paulo Madeiro
e-mail: jpaulo.vale@dc.ufc.br

J. A. Lobo Marques (✉)
Laboratory of Applied Neurosciences, University of Saint Joseph, Macao SAR, China
e-mail: alexandre.lobo@usj.edu.mo

J. A. Lobo Marques and S. J. Fong (eds.), *Computerized Systems for Diagnosis and Treatment of COVID-19*, https://doi.org/10.1007/978-3-031-30788-1_9

four combinations for classification scenarios are tested: (*Low vs. Moderate*, *Low vs. Severe*, *Moderate vs. Severe*) and 1 Multi-class comparison (*All vs. All*)). The results indicate that some statistically significant parameter distributions were found for all comparisons. (*Low vs. Moderate—Approximate Entropy* p-value = 0.0067 < 0.05, *Low vs. Severe—Correlation Dimension* p-value = 0.0087 < 0.05, *Moderate vs. Severe—Correlation Dimension* p-value = 0.0029 < 0.05, *All vs. All—Correlation Dimension* p-value = 0.0185 < 0.05. The non-linear analysis of the time-frequency representation of the ECG signal can be considered a promising tool for describing and distinguishing the COVID-19 severity activity along its different stages.

Keywords COVID-19 · ECG signals · Non-linear analysis · Statistical analysis

1 Introduction

In 2020, the World Health Organization (WHO) declared the Coronavirus Disease 19 (COVID-19) pandemic. Nowadays, with well over 6 million deaths worldwide [1], the scientific community is developing new ways to detect the disease and the pathology stages.

A Severe Acute Respiratory Syndrome-Coronavirus-2 causes COVID-19 (SARS-CoV-2) [2]. SARS-CoV-2 uses Angiotensin Converting Enzyme 2 (ACE2), a receptor in the cell surface, to initiate the infection process. The ACE2 are receptor present in the kidney, blood vessels, heart, and lungs, which means it can cause respiratory, cardiovascular, gastrointestinal, and central nervous system diseases [3–5].

The diagnosis of COVID-19 cases is a crucial step in controlling the disease. Laboratory testing techniques like Nucleic Acid Amplification Test (NAAT) and Antigen detection have proven to be accurate diagnosis techniques. NAAT is a technology that uses Real-Time Polymerase Chain Reaction (RT-PCR) assay to detect SARS-CoV-2. Antigen detection tests are used to detect the presence of viral proteins, in this case, SARS-CoV-2 [6].

L. Zacarias · C. Leite
Graduate Program in Cardiovascular Sciences, Federal University of Ceara,
Fortaleza, Ceara, Brazil
e-mail: camilafleite@ufc.br

M. Alves Neto · A. Aires Peixoto Jr
Department of Clinical Medicine, Faculty of Medicine, Federal University of Ceara, Fortaleza, Ceara, Brazil
e-mail: arnaldoapj@ufc.br

A. de Oliveira
University for the International Integration of the Afro-Brazilian Lusophony,
Fortaleza, Ceara, Brazil

The majority of the impact of COVID-19 is focused on the respiratory system; However, due to the nature of the virus, it can cause a variety of cardiac complications, with some studies showing the appearance of myocardial injury, heart failure, cardiogenic shock, and cardiac arrhythmias [7, 8].

The Eletrocardiogram (ECG) is an exam that monitors the heart's electrical activity by placing electrodes on the skin's surface and registers the actions potentials created by the cardiac cells in the form of electrical signals [9–11].

The ECG records waves created by the different moments of the cardiac cycle. The normal or expected cardiac cycle starts with the P wave, followed by the QRS complex, and ends with the T wave [9, 10].

While the detection of COVID-19 is an important step in controlling the disease. Pathology prognosis prediction reduces disease complications, and patient mortality [12, 13].

Multiple studies concluded that the visual representation of an abnormal T wave or the presence of S-T segment elevation/depression could have a good prognostic in predicting the mortality of COVID-19 patients [14–16], showing that the use of ECG signal to predict the severity might be an interesting alternative.

This research was conducted to evaluate the classification capacity of different non-linear features of the ECG signal, considering a transform to the time-frequency domain, to distinguish between 3 different degrees of severity or stages of COVID-19.

2 ECG Collection

The ECG signals were obtained following the recommendations of [17] to reduce circadian influences, and [18] for the correct signal stabilization, both for supine and orthostasis positions. The electrocardiographic recordings were performed using an Olimex EKG-EMG biosensor™cardiac monitor (OLIMEX (2014) SHIELD-EKG-EMG biofeedback shield USER'S MANUAL tech. rep.) of three channels connected to a microcomputer via Bluetooth technology. The frequency sample applied 256 Hz. The volunteers were monitored, in spontaneous breathing, in the modified MC5 lead.

3 Patients' Selection

The patients' selection considered both male and female hospitalized individuals, presenting clinical symptoms and diagnostic of COVID-19 confirmed by RT-PCR test. The patients were classified according to the severity shown in Table 1. The database is then divided into three different severity levels:

- low severity—eight patients
- moderate severity—nine patients
- severe—thirty three patients.

Table 1 COVID-19 severity classification criteria [19]

Low	Mild clinical symptoms and no signs of pneumonia on examination image
Moderate	Presence of fever and respiratory symptoms with evidence radiology of pneumonia
Severe	Respiratory distress (> 30 breaths/min)
	Oxygen saturation < 93% at rest
	Partial pressure of arterial oxygen (PaO_2)/fraction of inspired oxygen (FiO_2) <300 mmHg
	Cases with chest images that showed >50% evident lesion progression within 24–48 h
Critical	Respiratory failure and requiring mechanical ventilation
	Shock
	Other organ failures that require care in the ICU

Table 2 Socio-demographic and clinical characteristics

		Continuous measure	Min	Max		AVG	SD
Age			18	88		54.59	17.23
		Categorical Measure			%		
Gender	Male		60				
	Female		40				
Clinical Information	Low				16		
	Moderate				18		
	Severe				66		

The Ethics Committee approved the study for Research on Humans (CAAE—HELV: 47229221.9.3001.5684; CAAE—HUWC: 47229221.9.0000.5045).

During the analysis of the records, the data obtained during the time of postural changes due to signal instability were excluded.

Table 2 shows the Continuous Measure and Categorical Measure of the data present in the previously mentioned database.

4 Methodology

The proposed methodology is divided into five main steps:

1. Preprocessing/Artifacts Removal;
2. Windowing Signal Analysis Crops;
3. Feature Extraction;

4. Time-series Feature Distributions Data Compressor;
5. Statistical Analysis.

4.1 Artifacts Removal

The ECG signals raw data in the database showed some artifacts, and to ensure the signal quality, one of two actions have been performed per signal: (1) full signal deletion or (2) removal of the signal portions related to the artifacts if the rest of the signal has an acceptable quality. In the beginning, the database had signals from 51 patients. After the artifacts removal stage, the number of available signals in the database for the next stages was reduced to 42 patients (8 patients with low severity, six patients with moderate severity, and 27 patients with a severe stage of the disease).

4.2 Signal Analysis and Feature Extraction

For each study participant ECG recording, ten non-linear features have been extracted every 1 the s-long segment in 6 ways of windowing signal analysis crops: Rectangular non-overlapping windowing (Rec), Rectangular 0.5s overlapping windowing ($RecO$), Hamming non-overlapping windowing (Hm), Hamming 0.5s overlapping windowing (HmO), Hanning non-overlapping windowing (Hn) and Hanning 0.5s overlapping windowing (HnO). The ten extracted features were:

- **Energy** (En) of discrete signal $x(n)$ is defined as:

$$En = \sum_{n=0}^{N-1} |x(n)|^2 \tag{1}$$

  the En is calculated as the sum of each squared signal component, and it can be seen as the system's ability to accomplish work [20].
- **Entropy** is a measure that considers the amount of energy present in a complex system. Entropy features may also be used to quantify the information content that has been masked in a signal, modeling the unpredictability and irregularities of a pathological signal [20]. For this research, three different entropies are considered: Approximate (EA), Logarithmic ($ELog$), and Shannon ($ESha$). The three types of entropies can be defined as follows:

$$EA(m, r) = \lim_{N \to \infty} \Theta^m(r) - \Theta^{m+1}(r), \tag{2}$$

$$ELog = \sum_{n=1}^{N} \log[|x(n)|^2] \tag{3}$$

and

$$ESha = -\sum_{n=1}^{N} |x(n)|^2 \log[|x(n)|^2] \tag{4}$$

where N is the vector length, r denotes a tolerance value, Θ is the Heaviside step function and m is the dimension [21].

- **Hurst Exponent** (EH) is used as a measure of how chaotic or unpredictable a time series is [22]. The equation can be defined as:

$$K_q(\tau) \sim \left(\frac{\tau}{\nu}\right)^{qEH(q)}, \tag{5}$$

with

$$K_q(\tau) = \frac{\langle |X(t+\tau) - X(t))|^q \rangle}{\langle |X(t)|^q \rangle}, \tag{6}$$

where q is the order moments of the distribution increments, ν is the time resolution and t is the period of a given time series signal $X(t)$ [22].

- **Lyapunov Exponent** ($ELya$) is a measure of system's predictability and sensitivity for changing [23].

$$ELay(x_0) = \lim_{n\to\infty} \frac{\sum_{k=1}^{n} \ln|f'(x_k - 1)|}{n}, \tag{7}$$

where f' is the integrator function f derivative [24].

- **Higuchi Fractal Dimension** (H) estimates the fractal dimension of a time series signal. The equation can be illustrated as:

$$H = \frac{\ln(L(k))}{\ln(\frac{1}{k})}, \tag{8}$$

where k is a number of composed sub-series and $L(k)$ is the averaged curve length.

- **Katz Fractal Dimension** (K) estimates the fractal dimensions through a waveform analysis of a time series [22]. The equation is defined as:

$$K = \frac{\log(n)}{\log(n) + \log\left(\frac{max_n(\sqrt{(n-1)^2 + (X(n) - X(1))^2})}{\sum_{n=2}^{N}\sqrt{1 + (X(n-1) - X(n))^2}}\right)}, \tag{9}$$

where $X(n)$ is the signal.

- **Correlation Dimension** ($CorrDim$) is a measure to quantify self-similarity. A large $CorrDim$ means a high degree of complexity and less-similarity [21].

$$CorrDim = \lim_{M \to \infty} \frac{2 \sum_{i=1}^{M-k} \sum_{j=i+k}^{M} \Theta(l|X_i - X_j|)}{M^2}, \tag{10}$$

where $\Theta(x)$ is the Heaviside step function, the X_i and X_j are the position vectors on attractor, l is the distance under consideration, k is the summation offset and M is the reconstructed vectors numbers from the original signal [21].

- **Detrended Fluctuation Analysis** (DFA) is a technique for measuring the power scaling observed through R/S analysis and it's specifically used to address non-stationary time-series [25]. The DFA can be calculated as

$$DFA(n) = \sqrt{\frac{\sum_{k=1}^{N}[y(k) - y_n(k)]^2}{N}}, \tag{11}$$

where n is the length, $y_n(k)$ is the local trend and $y(k)$ is define as

$$y(k) = \sum_{i=1}^{k}[x(i) - \overline{x}], \tag{12}$$

with $x(i)$ being the interbeat interval and $\overline{x}$ being the average interbeat interval [26].

4.3 Feature Time-Series Data Vector Compression

Treating the ten extracted features of all segments as time series distributions per different windowing analyses, six different statistical functions act like data information compressors over time, reducing the problem dimensionality and, at the same time, ensuring that the number of metrics per subject was equally balanced for statistical analysis task. The six statistical functions were,

- **Average** (Avg) is calculated by the sum of values in the vector and then divided by the length of the vector [27]. The equation is defined as

$$Avg = \frac{\sum_{n=1}^{N} x(n)}{N} \tag{13}$$

where N is the length of the vector and x is the vector.

- **Standard Deviation** (Std) is a measure of how dispersed the data is relative to the mean [27] and can be defined as

$$Std = \sqrt{\frac{\sum_{n=1}^{N} |x(n) - Avg|^2}{N - 1}} \tag{14}$$

- **95th Percentile** ($P95$) is the number that is grater than 95% of the values in the vector. $P95$ is obtained by ordering the values in ascending order and using the formula

$$n_{P95} = 0.95 * K, \tag{15}$$

where n is the $P95$ position in a vector and K is the number of samples.
- **Variance** (Var) is the spread of numbers in a vector and can be defined as

$$Var = \frac{\sum_{n=1}^{N} |x(n) - Avg|^2}{N - 1} \tag{16}$$

- **Median** (Med) is the value separating the highest values in half from the lowest values. Med is obtained by arranging the values in ascending order and using the equation

$$n_{Med} = 0.5 * K, \tag{17}$$

where n_{Med} is the Med position in a array and K is the number of samples.
- **Kurtosis** (Kur) measures how a distribution differs from a normal distribution. The equation is defined as

$$Kur = \frac{\sum_{n=1}^{N} |x(n) - Avg|^4}{Std^4}. \tag{18}$$

4.4 Data Normalization

Before entering the statistical analysis, the data went through a data normalization process, for this research we used the $z - score$ normalization. The $z - score$ is a function that measures the distance of a certain data point time-series from its Avg according to its Std [27]. The $z - score$ equation can be defined as

$$z - score = \frac{X - Avg}{Std}, \tag{19}$$

where X is a data point time-series distribution.

Each time-series feature vector per position, per windowing analysis crop, per study group comparative analysis (*Low vs. Moderate*, *Low vs. Severe*, *Moderate vs. Severe*, *All vs. All*) have been suffered a normalization by $z - score$.

4.5 Statistical Analysis

In this study, the searches for statistical differences were performed for each pair of study groups—3 Binary comparison (*Low vs. Moderate*, *Low vs. Severe*, *Moderate vs. Severe*) and 1 Multi-class comparison (*All vs. All*).

The normality and homoscedasticity of the time-series feature vector distributions have been assessed with the Kolmogorov-Smirnov and Levene tests, respectively. The hypothesis of parametric tests was not met. Therefore, the non-parametric tests have been applied for the proposed: Kruskal-Wallis test with the Bonferroni multi-classes correction for the comparison *All vs. All* and Mann-Whitney test for the 3 Binary comparison groups.

The total number of statistical tests done were 2880, resulting from 60 time-series features distributions per 4 comparative analyses, per 6 windowing signal analysis crops, and per 2 patient positions data acquisitions.

For a better understanding of the whole methodology, Fig. 1 presents a summary of the procedure.

5 Results

In this section, we present the best p-values achieved in the statistical analysis stage per feature time-series analysis for each study group comparison (Tables 3, 4, 5 and 6). The results selection is made per windowing signal analysis crop and per data compressor in two different body positions. The N.S. means not significant. The Box-plots of the time-series feature distributions (normalized by z-score) that present significant differences in each study group comparison are illustrated in Figs. 2, 3, 4 and 5. In each boxplot figure the feature time-series distribution is labeled as Position_Feature_Window_Statistical, and the statistical achieved p-value is presented on top of the figure box. The ordered data distribution values in these diagrams have been divided into four parts using a box with three horizontal lines. The distribution sample median was displayed as a line lying approximately in the middle of the box, where the 25th and 75th percentiles of the sample fall at the bottom and top of the box, respectively. Finally, two dashed discrepancy lines with marks are the whiskers located on the last box that extends to a maximum of 1.5 times the interquartile range. The values beyond these lines are marked by the "+" symbol as ultra discrepant data.

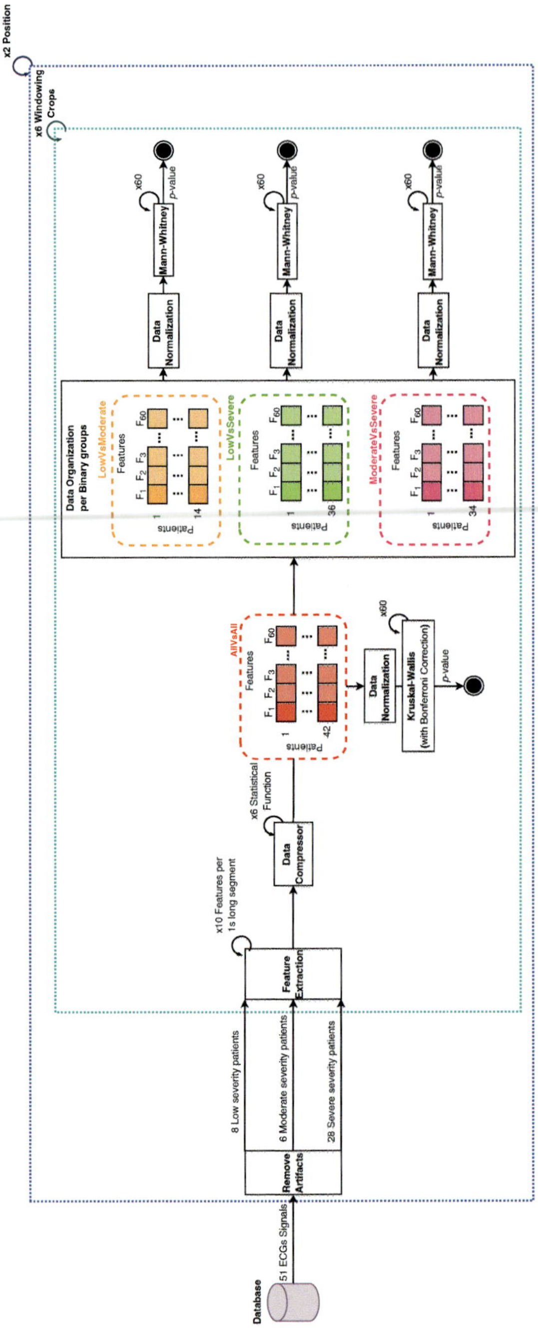

Fig. 1 Methodology workflow diagram

Table 3 Mann-Whitney statistical test– -*Low versus Moderate* binary comparison

Position	Features	Window	Statistical	p-value
Down	CorrDim	Rec	Kur	**0.0098**
	DFA	Rec	Med	N.S.
	En	Hn	P95	N.S.
	EA	Rec	Std	N.S.
	ELog	HmO	Var	N.S.
	ESha	HmO	Med	N.S.
	EH	Hm	Std	N.S.
	ELay	Hn	Std	N.S.
	H	Rec	Std	N.S.
	K	Hm	P95	N.S.
Up	CorrDim	Hn	Std	N.S.
	DFA	HmO	Avg	N.S.
	En	HmO	Avg	N.S.
	EA	HmO	Kur	**0.0067**
	ELog	HmO	Kur	N.S.
	ESha	RecO	Var	N.S.
	EH	HnO	P95	N.S.
	ELay	Hn	Avg	N.S.
	H	Hm	Kur	**0.0389**
	K	HmO	Avg	N.S.

6 Discussion

Looking into the Binary comparison between *Low verses Moderate*, present in Table 3, we can see that 3 features showed significant differences (p-values < 0.05), two in the *Up* position and one in the *Down* position. The feature EA showed the best p-value $= 0.0067 < 0.05$ with Kur as its time-series data compressor and HmO as a windowing data analysis crop.

In the Binary comparison between *Low verses Severe*, visualized in Table 4, 5 features showed significant differences, 3 of them come from the *Up* position. The highest significant time-series features distributions provided a p-value $= 0.0086 < 0.05$ ($CorrDim$ and $ELog$, with Std and Kur as data compressors and Hn and Hm as windowing data analysis crops, respectively).

The Table 5, which is related to the comparison between the *Moderate vs. Severe*, has 6 time-series feature distributions that showed significant differences, 4 of them located in *Down* position and 2 in the *Up* position. The best achieved p-value $= 0.0029 < 0.05$, a significant difference attained by Kur compression of the $CorrDim$ feature in the $RecO$ windowing data analysis crop.

Table 6, which is related to the *All vs. All* comparison group, has 2 feature distributions that showed significant differences, one for each position. The best achieved

Table 4 Mann-Whitney statistical test—*Low verses Severe* binary comparison

Position	Features	Window	Statistical	*p*-value
Down	CorrDim	Hm	Med	**0.0364**
	DFA	RecO	Std	N.S.
	En	HnO	Kur	N.S.
	EA	Rec	Med	N.S.
	ELog	HmO	Kur	N.S.
	ESha	Hn	P95	N.S.
	EH	Hn	Kur	**0.0224**
	ELay	HnO	Med	N.S.
	H	HmO	Std	N.S.
	K	RecO	Kur	N.S.
Up	CorrDim	Hn	Std	**0.0087**
	DFA	HmO	Med	N.S.
	En	RecO	Med	N.S.
	EA	Hm	Med	N.S.
	ELog	Hm	Kur	**0.0086**
	ESha	Rec	P95	N.S.
	EH	Hn	P95	N.S.
	ELay	RecO	Avg	N.S.
	H	Hn	Kur	**0.0399**
	K	Rec	Med	N.S.

Table 5 Mann-Whitney statistical test—*Moderate verses Severe* binary comparison

Position	Features	Window	Statistical	*p*-value
Down	CorrDim	RecO	Kur	**0.0029**
	DFA	HmO	Avg	**0.0377**
	En	HmO	Var	**0.0448**
	EA	Rec	Std	N.S.
	ELog	Hm	Std	N.S.
	ESha	RecO	Avg	**0.0421**
	EH	Hn	Std	N.S.
	ELay	Rec	P95	N.S.
	H	Rec	Kur	N.S.
	K	HmO	Std	N.S.
Up	CorrDim	HmO	Std	N.S.
	DFA	HmO	Avg	N.S.
	En	Hm	Avg	N.S.
	EA	Rec	Std	**0.0239**
	ELog	Hm	P95	N.S.
	ESha	Hn	Avg	N.S.
	EH	Hm	P95	N.S.
	ELay	Hn	Std	N.S.
	H	HmO	Std	**0.0188**
	K	Hm	Avg	N.S.

Table 6 Kruskall-Wallis statistical test—*All verses All* comparison

Position	Features	Window	Statistical	*p*-value
Down	CorrDim	Rec	Kur	**0.0185**
	DFA	HnO	P95	N.S.
	En	Hn	P95	N.S.
	EA	RecO	Med	N.S.
	ELog	HmO	Kur	N.S.
	ESha	Hn	P95	N.S.
	EH	Hn	Kur	N.S.
	ELay	HmO	Kur	N.S.
	H	Rec	Kur	N.S.
	K	HmO	Kur	N.S.
Up	CorrDim	RecO	Avg	N.S.
	DFA	RecO	Kur	N.S.
	En	Hm	Kur	N.S.
	EA	HnO	Kur	**0.0477**
	ELog	HmO	Kur	N.S.
	ESha	Hm	Kur	N.S.
	EH	Hn	P95	N.S.
	ELay	HmO	Std	N.S.
	H	HmO	Std	N.S.
	K	Hm	Kur	N.S.

p-value = 0.0185 < 0.05, a significant difference attained by Med data compression of the $CorrDim$ feature in the HmO windowing data analysis crop.

Now looking into all tables at the same time (Tables 3, 4, 5 and 6), the position that provided the most significant time-series feature distributions was the *Up* position. The time-series distributions related to $CorrDim$ had the highest quantity of p-values < 0.05. The statistical function data compressor that provided a more significant time-series feature distribution was Kur with 7 p-values < 0.05. The best Windowing data analysis crop process was HmO with 5 p-values bellow 5%.

Globally, *Up* position, Rec and Hn windowing signal analysis crops, the feature $CorrDim$, and the data compressor Kur gave more quality information to separate binary classes than other modalities.

When looking into the 3 different classes, we could see that the *Moderate* class typically presented parameter distributions with a Median value above all other classes. For the *Low* class, the Median of all feature distributions was typically lower than the other classes except the one illustrated in subfigure Fig. 3d, where the Median value is slightly higher than the *Severe* class Median.

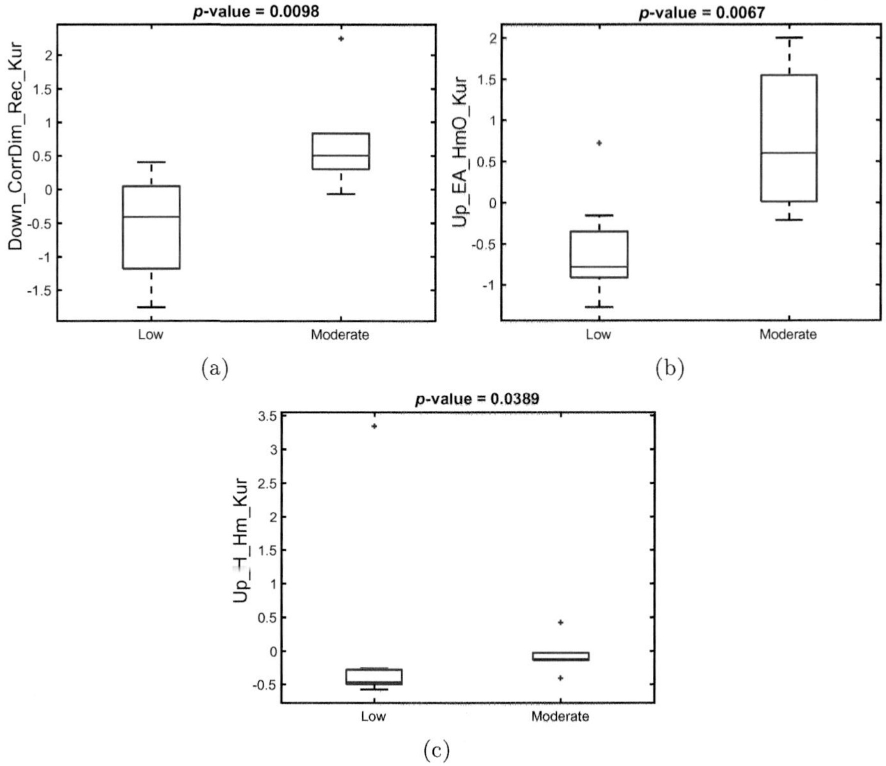

Fig. 2 *Low verses Moderate* Binary comparison boxplots of the distributions of: **a** Down_CorrDim_Rec_Kur; **b** Up_EA_HmO_Kur; **c** Up_H_Hm_Kur

7 Conclusion

On this investigation 10 non-linear features (*Energy*, *Approximate Entropy*, *Logarithmic Entropy*, *Shannon Entropy*, *Hurst Exponent*, *Lyapunov Exponent*, *Higuchi Fractal Dimension*, *Katz Fractal Dimension*, *Correlation Dimension* and *Detrended Fluctuation Analysis*) extracted from 2 ECG signals, collected from 2 different patient's locations, every 1 the s-long segment in 6 ways of windowing signal analysis crops were evaluated employing statistical analysis to find differences between COVID-19 evolution stages (*Low vs. Moderate*, *Low vs. Severe*, *Moderate vs. Severe*) and 1 Multi-class comparison (*All vs. All*)). The results showed that a non-linear analysis could be considered a good tool for describing and distinguishing the COVID-19 ECG activity along its different stages as some statistically significant parameter distributions were found for all comparisons (*Low vs. Moderate—Approximate Entropy* p-value $= 0.0067 < 0.05$, *Low vs. Severe—Correlation Dimension* p-value $= 0.0087 < 0.05$, *Moderate vs. Severe—Correlation Dimension* p-value $= 0.0029 < 0.05$, *All vs. All—Correlation Dimension* p-value $= 0.0185 < 0.05$).

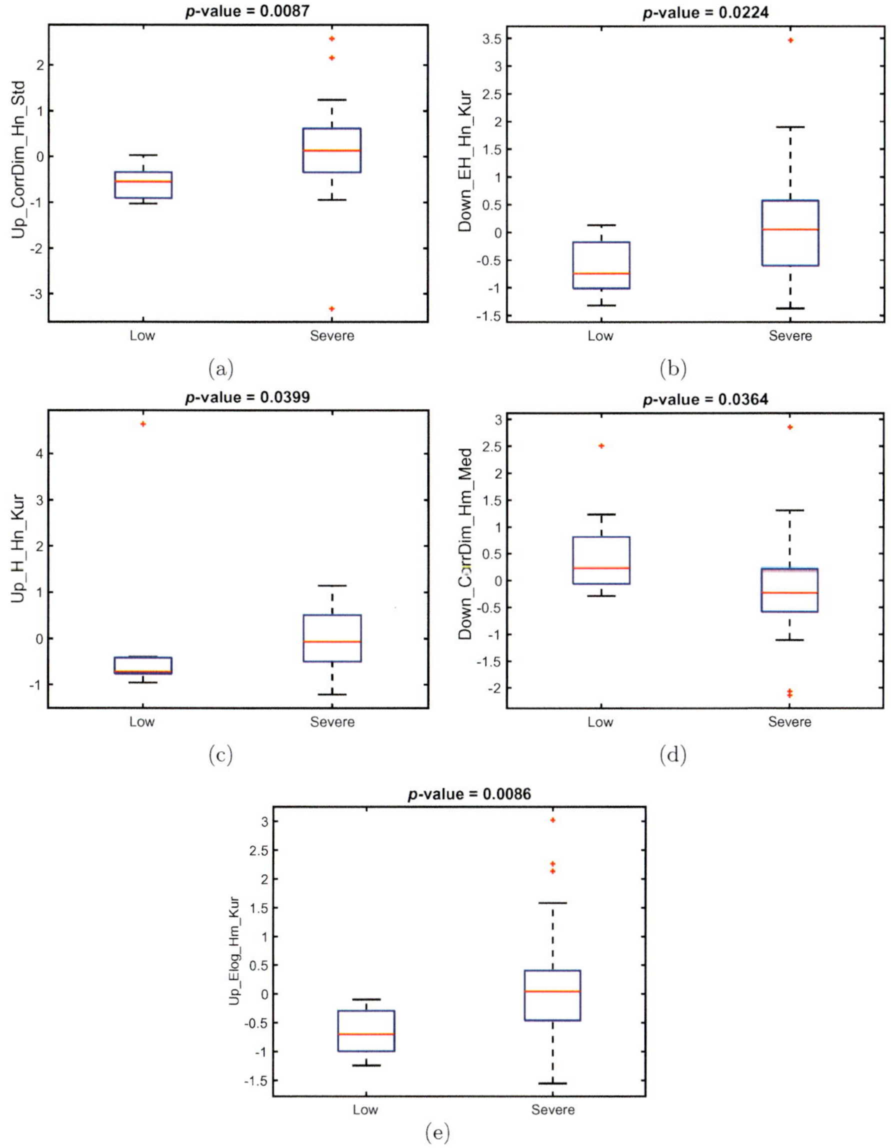

Fig. 3 *Low verses Severe* binary comparison boxplots of the distributions of: **a** Down_EH_Hn_Kur; **b** Up_CorrDim_Hn_Std; **c** Up_H_Hn_Kur; **d** Down_CorrDim_Hm_Med; **e** Up_Elog_Hm_Kur

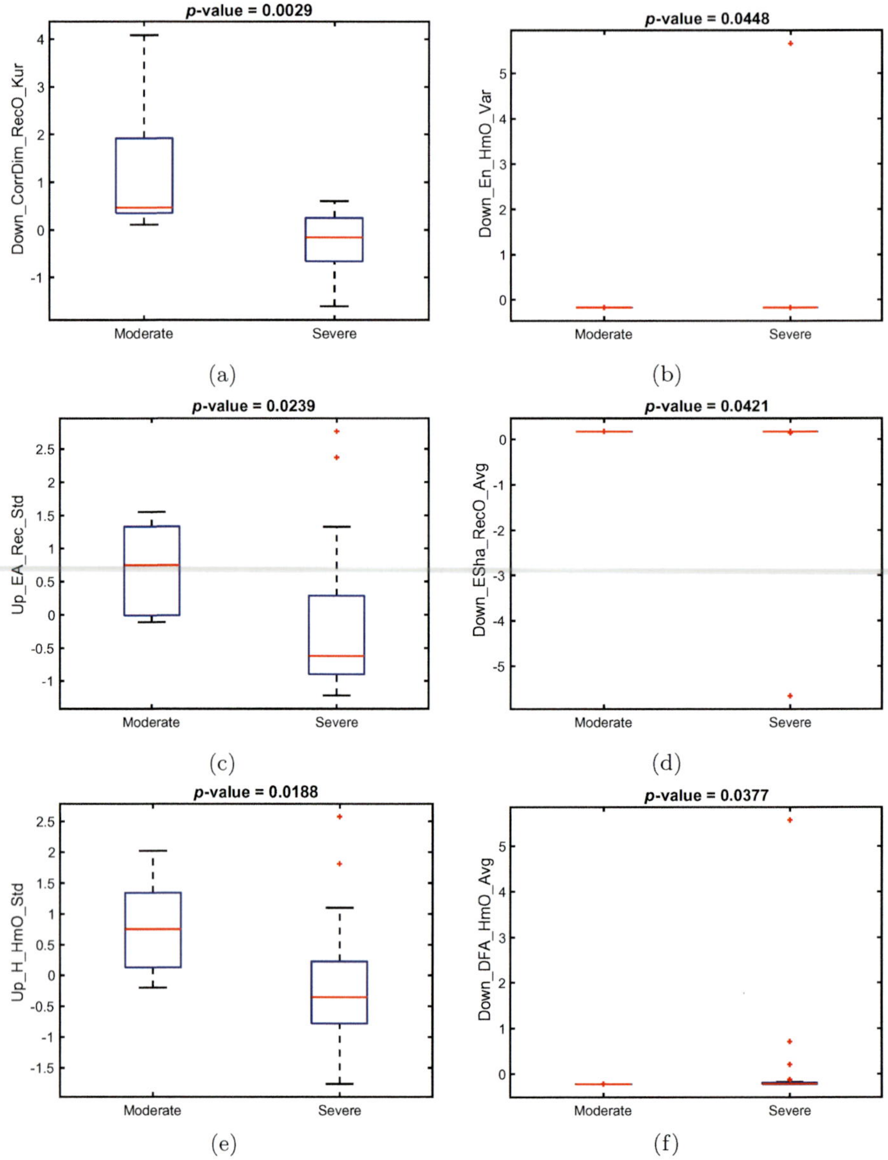

Fig. 4 *Moderate verses Severe* binary comparison boxplots of the distributions of: **a** Down_DFA_HmO_Avg; **b** Down_En_HmO_Var; **c** Up_H_HmO_Std; **d** Down_CorrDim_RecO_Kur; **e** Down_ESha_RecO_Avg; **f** Up_EA_Rec_Std

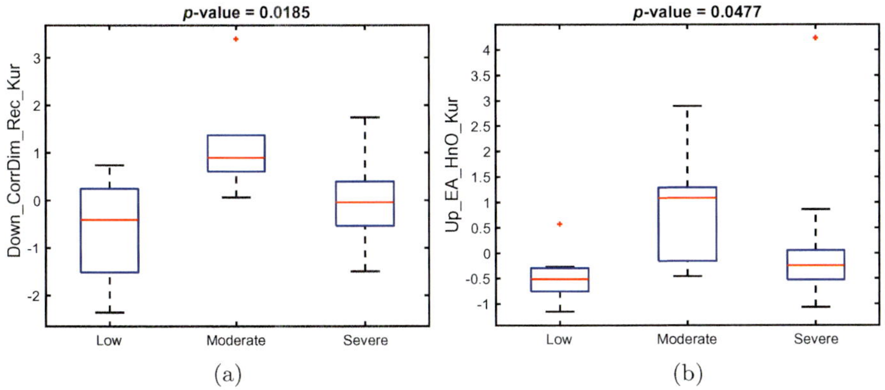

Fig. 5 *All verses All* comparison boxplots of the distributions of: **a** Down_CorrDim_Rec_Kur; **b** Up_EA_HnO_Kur

In future research will be interesting to use and combine the significant time-series features found in this study as inputs for Machine Learning models to verify the possibility of creating a system for COVID-19 diagnostic and prognostic.

Acknowledgements This work was supported by:
National Funds from FCT—Fundação para a Ciência e a Tecnologia through project UIDB/50016/2020.
Funcap (Ceara State Foundation for the Support of Scientific and Technological Development) Grant PS1-0186-00439.01.00/21.

References

1. Who coronavirus (covid-19) dashboard. https://covid19.who.int/. Accessed from 22 Aug 2022
2. Mehraeen E, Alinaghi SAS, Nowroozi A, Dadras O, Alilou S, Shobeiri P, Behnezhad F, Karimi A (2020) A systematic review of ECG findings in patients with COVID-19. Indian Heart J 72(6):500–507 Nov
3. Yesudhas D, Srivastava A, Michael Gromiha M (2020) COVID-19 outbreak: history, mechanism, transmission, structural studies and therapeutics. Infection 49(2):199–213
4. Hosseini ES, Kashani NR, Nikzad H, Azadbakht J, Bafrani HH, Kashani HH (2020) The novel coronavirus disease-2019 (COVID-19): mechanism of action, detection and recent therapeutic strategies. Virology 551:1–9
5. Ciotti M, Ciccozzi M, Terrinoni A, Jiang W-C, Wang C-B, Bernardini S (2020) The COVID-19 pandemic. Crit Rev Clin Lab Sci 57(6):365–388 Jul
6. Lai CKC, Lam. W (2021) Laboratory testing for the diagnosis of COVID-19. Biochem Biophys Res Commun 538:226–230
7. Long B, Brady WJ, Bridwell RE, Ramzy M, Montrief T, Singh M, Gottlieb M (2021) Electrocardiographic manifestations of COVID-19. Am J Emerg Med 41:96–103 Mar
8. Farshidfar F, Koleini N, Ardehali H (2021) Cardiovascular complications of covid-19. JCI insight 6(13)
9. Thaler M (2018) Only ECG book You'll ever need 9th. Wolters Kluwer Health

10. Strauss DG, Wagner GS (2013) Marriott's practical electrocardiography. Lippincott Williams & Wilkins
11. John R (2013) Hampton. The ECG made easy, Churchill Livingstone/Elsevier
12. Booth AL, Abels E, McCaffrey P (2020) Development of a prognostic model for mortality in COVID-19 infection using machine learning. Mod Pathol 34(3):522–531 Oct
13. Subudhi S, Verma A, Patel AB (2020) Prognostic machine learning models for COVID-19 to facilitate decision making. Int J Clin Pract 74(12)
14. Aghajani MH, Toloui A, Aghamohammadi M, Pourhoseingholi A, Taherpour N, Sistanizad M, Neishaboori AM, Asadpoordezaki Z, Miri R (2021) Electrocardiographic findings and in-hospital mortality of covid-19 patients; a retrospective cohort study. Arch Acad Emerg Med 9:e45
15. Kaliyaperumal D, Bhargavi K, Ramaraju K, Nair KS, Ramalingam S, Alagesan M (2022) Electrocardiographic changes in COVID-19 patients: a hospital-based descriptive study. Indian J Crit Care Med 26(1):43–48
16. Wang Y, Chen L, Wang J, He X, Huang F, Chen J, Yang X (2020) Electrocardiogram analysis of patients with different types of COVID-19. Ann Noninvasive Electrocardiol 25(6)
17. Dietrich DF, Schindler C, Schwartz J, Barthélémy J-C, Tschopp J-M, Roche F, von Eckardstein A, Brändli O, Leuenberger P, Gold DR, Gaspoz J-M, Ackermann-Liebrich U (2006) Heart rate variability in an ageing population and its association with lifestyle and cardiovascular risk factors: results of the SAPALDIA study. EP Europace 8(7):521–529
18. Brown SJ, Brown JA (2007) Resting and postexercise cardiac autonomic control in trained masters athletes. J Physiol Sci 57(1):23–29
19. Yan X, Han X, Peng D, Fan Y, Fang Z, Long D, Xie Y, Zhu S, Chen F, Lin W, Zhu Y (2020) Clinical characteristics and prognosis of 218 patients with COVID-19: a retrospective study based on clinical classification. Front Med 7
20. Sundararajan D (2015) Discrete wavelet transform a signal processing approach. Wiley
21. Caesarendra W, Kosasih B, Tieu K, Moodie CAS (2013) An application of nonlinear feature extraction - a case study for low speed slewing bearing condition monitoring and prognosis. In: 2013 IEEE/ASME international conference on advanced intelligent mechatronics. IEEE
22. Garcia AAT, Garcia CAR, Villasenor-Pineda L, Montoya OM (2022) Biosignal processing and classification using computational learning and intelligence principles, algorithms, and applications. Academic
23. Rüdisüli M, Schildhauer TJ, Biollaz SMA, Van Ommen JR (2013) Measurement, monitoring and control of fluidized bed combustion and gasification. In: Fluidized bed technologies for near-zero emission combustion and gasification. Elsevier, pp 813–864
24. Silva G, Batista P, Rodrigues PM (2022) COVID-19 activity screening by a smart-data-driven multi-band voice analysis. J Voice
25. Bryce RM, Sprague KB (2012) Revisiting detrended fluctuation analysis. Sci Rep 2(1)
26. Hardstone R, Poil S-S, Schiavone G, Jansen R, Nikulin VV, Mansvelder D, Linkenkaer-Hansen K (2012) Detrended fluctuation analysis: a scale-free view on neuronal oscillations. Front Physiol 3
27. Peck R, Olsen C, Devore JL (2008) Introduction to statistics and data analysis. Cengage Learning

Classification of Severity of COVID-19 Patients Based on the Heart Rate Variability

Daniel Pordeus, Pedro Ribeiro, Laíla Zacarias, João Paulo Madeiro, Joao Alexandre Lobo Marques, Pedro Miguel Rodrigues, Camila Leite, Manoel Alves Neto, Arnaldo Aires Peixoto Jr, and Adriel de Oliveira

Abstract The continuous development of robust machine learning algorithms in recent years has helped to improve the solutions of many studies in many fields of medicine, rapid diagnosis and detection of high-risk patients with poor prognosis as the coronavirus disease 2019 (COVID-19) spreads globally, and also early prevention of patients and optimization of medical resources. Here, we propose a fully

D. Pordeus · J. Paulo Madeiro
Department of Computing, Federal University of Ceará, Fortaleza, Ceará, Brazil
e-mail: pordeus@alu.ufc.br

J. Paulo Madeiro
e-mail: jpaulo.vale@dc.ufc.br

P. Ribeiro · P. Miguel Rodrigues
CBQF–Centro de Biotecnologia e Química Fina–Laboratório Associado, Escola Superior de Biotecnologia, Universidade Católica Portuguesa, Rua de Diogo Botelho 1327, 4169-005 Porto, Portugal
e-mail: s-pmsbribeiro@ucp.pt

P. Miguel Rodrigues
e-mail: pmrodrigues@ucp.pt

J. A. Lobo Marques (✉)
Laboratory of Applied Neurosciences, University of Saint Joseph, Macao SAR, China
e-mail: alexandre.lobo@usj.edu.mo

L. Zacarias · C. Leite
Graduate Program in Cardiovascular Sciences, Federal University of Ceara, Fortaleza, Ceara, Brazil
e-mail: camilafleite@ufc.br

M. Alves Neto · A. Aires Peixoto Jr
Department of Clinical Medicine, Faculty of Medicine, Federal University of Ceara, Fortaleza, Ceara, Brazil
e-mail: arnaldoapj@ufc.br

A. de Oliveira
University for the International Integration of the Afro-Brazilian Lusophony, Fortaleza, Ceara, Brazil

J. A. Lobo Marques and S. J. Fong (eds.), *Computerized Systems for Diagnosis and Treatment of COVID-19*, https://doi.org/10.1007/978-3-031-30788-1_10

automated machine learning system to classify the severity of COVID-19 from electrocardiogram (ECG) signals. We retrospectively collected 100 5-minute ECGs from 50 patients in two different positions, upright and supine. We processed the surface ECG to obtain QRS complexes and HRV indices for RR series, including a total of 43 features. We compared 19 machine learning classification algorithms that yielded different approaches explained in a methodology session.

Keywords Electrocardiogram (ECG) signal · Heart Rate Variability (HRV) indices · Signal processing · Severity · COVID-19

1 Introduction

In December 2019, a novel coronavirus was first recognized in Wuhan, China. On 30 January 2020, as infection rates and deaths across China soared and the first death outside China was recorded, the World Health Organization (WHO) described the then-unnamed disease as a Public Health Emergency of International Concern [1]. The disease was officially named coronavirus disease 2019 (COVID-19) by the WHO on 11 February 2020 and was declared a pandemic on 11 March 2020.

The coronavirus disease 2019 (COVID-19) has spread globally, and medical resources become insufficient in many regions. This condition can be diagnosed using RT-PCR technique on nasopharyngeal and throat swabs with sensitivity values ranging from 30 to 70%. However, chest CT scans and X-ray images have been reported to have sensitivity values of 98 and 69%, respectively, but, eventually, RT-PCR takes too long to respond to COVID-19 treatment.

Diagnosing the active cases of COVID-19 came as one of the most essential parts of controlling the pandemic. The accurate diagnosis of the pathology came from a laboratory testing technique named Nucleic Acid Amplification Test (NAAT). The method diagnoses an active COVID-19 infection in the upper respiratory tract using Real-Time Polymerase Chain Reaction(RT-PCR) assay [2, 3].

Since the appearance of the disease, the scientific community has been looking for different ways to identify the condition more quickly since time is vital to the success of the treatment. Several studies have addressed different possibilities of diagnosis approaches, most of them considering medical imaging, such as chest x-ray images, like [4, 5].

Since, there is significant variability of symptoms and side effects among COVID-19 patients, multiple approaches for diagnosis and treatment are under consideration by the medical and scientific communities, in addition to laboratory-based tests and medical imaging. For example, specialists are considering other approaches for groups of patients developing persistent symptoms such as fatigue, cardiac arrhythmia and neurological cognitive impairment. Examples are polysomnographic examinations and the monitoring of the Autonomous Nervous System (ANS) through the analysis of Heart Rate Variability.

Heart rate variability (HRV) is the analysis of the changes in the duration of intervals between adjacent heartbeats. It is an efficient technique for assessing interdependent regulatory systems that operate on different time scales to adapt to challenges and achieve optimal performance, as described in [6]. The work [6] review the balance between the sympathetic and parasympathetic subsystems in the heart and examine the interpretation of HRV, considering criteria for the association between reduced HRV and the risk of disease and mortality and the loss of regulatory capacity. This work also discusses the intrinsic cardiac nervous system and the heart-brain connection, through which afferent information can influence activity in the subcortical and frontocortical areas and motor cortex. It also highlights new perspectives on the underlying physiological mechanisms and properties of the ultra-low-frequency (ULF), very-low-frequency (VLF), low-frequency (LF), and high-frequency (HF) bands. In addition, it reviews the most common time and frequency domain measurements as well as standardized data collection protocols. The irregular behavior of the heartbeat is readily apparent when HR is examined on a beat-to-beat basis but is overlooked when a mean value over time is calculated. These fluctuations in HR result from complex, nonlinear interactions among several physiological systems. HRV is thus considered a measure of neurocardiovascular function that reflects heart-brain interactions and autonomic nervous system (ANS) dynamics [7].

We propose a solution for severity classification within patients diagnosed with Covid-19 using frequency-domain, time-domain and non-linear metrics extracted from the HRV signal to serve as a feature vector for machine learning algorithms. As a starting point, the severity classification originally considered 4 classes: low, moderate, severe and critical. Additional details can be found in Sect. 2. Due to the significant unbalance between classes, two groups of classes were proposed. The first one combines low and moderate patients, while the second group combines severe with critical. As the best results, we obtained an accuracy of 71% with the LinearSVC and F1-score as 1.00 with SVC and QuadraticDiscriminantAnalysis.

This work is organized as follows: Sect. 2 describes how the ECG dataset was obtained. Section 3 presents the technique used to extract the HRV metrics which make up the feature vector (input) for the machine learning classifier algorithms. Session 4 presents the applied training strategies, with the corresponding results, while Sessions 5 and 6 bring, respectively, Discussion and Conclusion.

2 Dataset

2.1 Dataset Creation

The inclusion criteria for this research considered hospitalized individuals of both genders, with clinical symptoms and diagnosis of COVID-19 confirmed by RT-PCR. The patients were classified according to the severity shown in Table 1. The database

Table 1 COVID-19 severity classification criteria [8]

Low	Mild clinical symptoms and no signs of pneumonia on examination image
Moderate	Presence of fever and respiratory symptoms with evidence radiology of pneumonia
Severe	Respiratory distress (>30 breaths/min)
	Oxygen saturation <93% at rest
	Partial pressure of arterial oxygen (PaO2)/ fraction of inspired oxygen (FiO2) <300 mmHg
	Cases with chest images that showed >50% evident lesion progression within 24–48 h
Critical	Respiratory failure and requiring mechanical ventilation
	Shock
	Other organ failures that require care in the ICU

is then divided into three different severity levels: 8 patients with low severity, 9 patients with moderate severity and 33 patients with a severe stage of the disease.

The ECG signals were obtained following the recommendations of [9] to reduce circadian influences and [10] for the correct signal stabilization, both for supine and orthostasis positions. The electrocardiographic recordings were performed using an Olimex EKG-EMG biosensor™cardiac monitor (OLIMEX (2014) SHIELD-EKG-EMG biofeedback shield USER'S MANUAL tech. rep.) of three channels connected to a microcomputer via Bluetooth technology. The frequency sample applied 256 Hz. The volunteers were monitored, in spontaneous breathing, in the modified MC5 lead.

3 ECG Data Processing

The general overview of the proposed data analysis is presented in Fig. 1.

The phases are described in the following subsections.

3.1 ECG Acquisition

The acquisition procedure consisted of acquiring ECG signals by an OLIMEX ECG biosensor (sampling frequency 256 Hz) with the following protocol: the patient remains lying down for five minutes (upright position) and then in the supine position for five minutes.

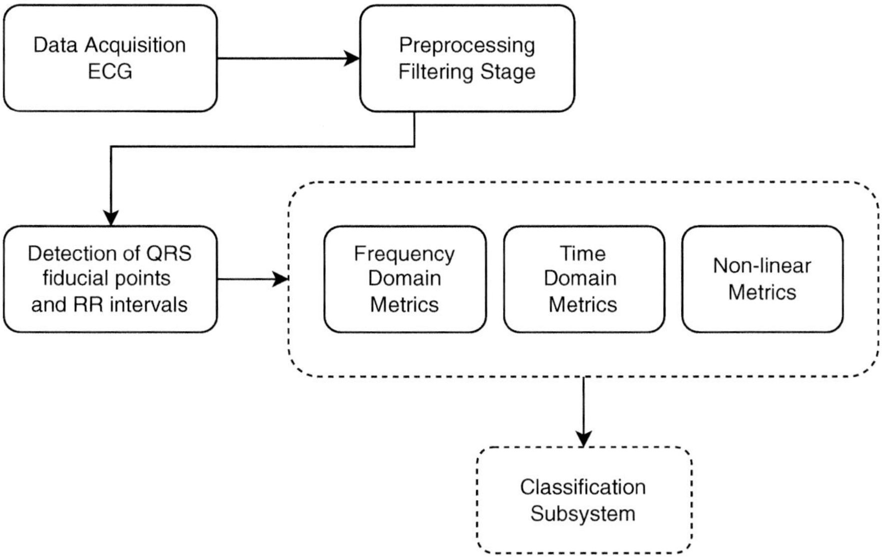

Fig. 1 General methodology for the Heart Rate Variability (HRV) analyzes in the context of individuals infected by COVID-19

3.2 *Preprocessing/Filtering Stage*

Due to the presence of multiple sources of noise and interference, discrete wavelet transform (DWT) is applied to suppress high-frequency and low-frequency noise. DWT can be applied using filter banks and downsampling and upsampling processes, as shown in the following diagram of Fig. 2.

Therefore, Daubechies10 function ("db10") [11] is used as the mother-wavelet which provides eliminating the coefficients resulting from high-pass filtering after the first decomposition stage. Then, the discrete wavelet transform (DWT) is used to remove low-frequency noise (baseline drift), now using the Daubechies4 function ('db4) with eight stages of decomposition and reconstruction. The coefficients

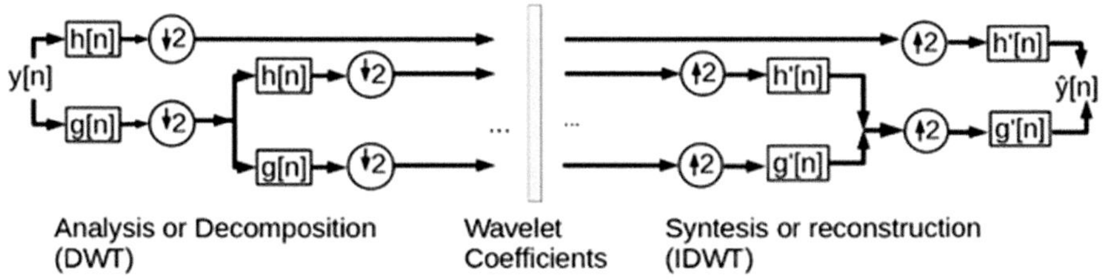

Fig. 2 Downsampling and upsampling processes

computed after the eighth level of decomposition are subtract the original signal after reconstruction process.

Data Cleaning

One of the ECG scan for two volunteers was faulty, which meant we had to remove that information from the collection. Furthermore, when running the method described in this Section only three NaN values was obtained. For these few cases we average the measures and fill in the gaps.

3.3 QRS Detection

The ECG signal generated by the filtering process can then be used in the QRS complex detection. Figure 3 illustrates how to obtain a time series of R-R intervals. The methods include a combination of continuous wavelet transform, Hilbert transform and derived filter techniques [12]. First, the signal is divided into 10-second windows, and we apply a feed-forward convolution to each window using the Mexican Hat function [13], taking into account two possible scaling factors. The first derivative is computed from the numerical method of first-order division. Finally, the Hilbert transform is computed for the resulting signal by taking the modulus of the resulting analytic signal.

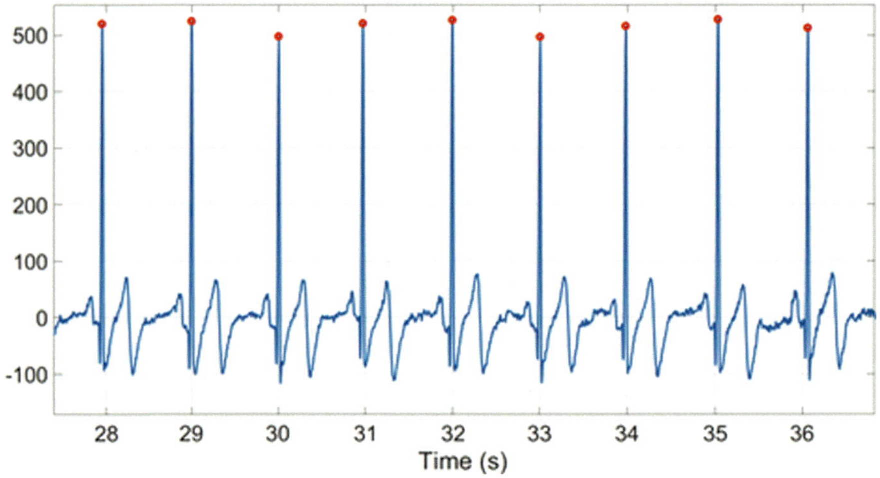

Fig. 3 Detected QRS complexes

RR Series

For each wavelet transform scaling factor applied, we have determined a detection threshold that allows detection of an acceptable number of beats, varying between 0.7 and 3.5 s of the window size. Therefore, the scaling factor is chosen such that beat detection in the filtered signal is obtained such that the product of the amplitude and the standard deviation of the interval is as small as possible. Figure 4 illustrate an example of signal collected using correctly detected beats and generated RR series.

Once the R-R signal generated by the QRS composite detection process was obtained, a moving average process was applied to a window series of 61 samples. Then we determine the first sample of the R-R series whose energy (amplitude squared) differs by no more than 15% from the energy of the corresponding sample of the determined moving average series. This sample is considered to be the first useful sample of the R-R series. In Fig. 5, we observe the R-R series and the corresponding moving average.

After the first useful sample is identified, the process of scanning the R-R time series begins to identify abnormal intervals associated with detection errors or ectopic beats, which are subsequently discarded and replaced by interpolation techniques. We assume that the kth sample of the RR series is the first useful sample of the series identified in the previous step, which we call the reference sample *RR_ref*:

$$RR_{ref} \leftarrow RR[k]$$

We scan from the next sample, RR[k+1]. If the difference between RR_ref and RR[k+1] is less than or equal to 160 ms, the reference pattern is updated to the value of RR[k+1], which means that pattern in the RR series is considered valid. In this In

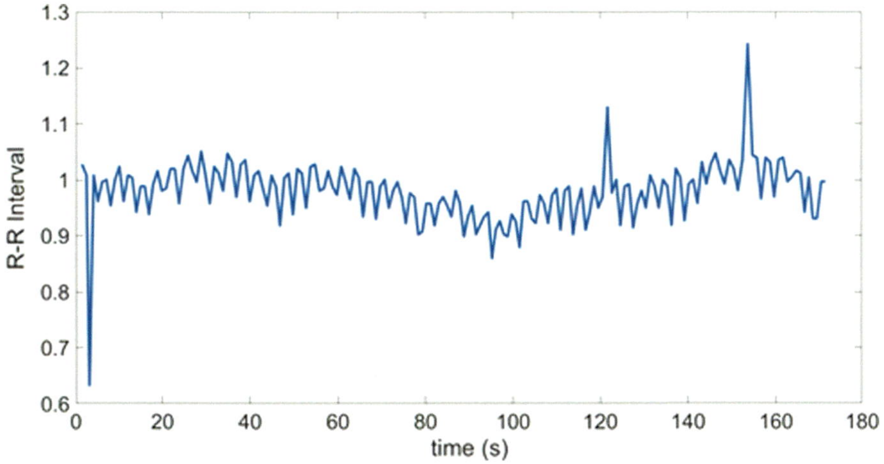

Fig. 4 Extracted R-R interval series

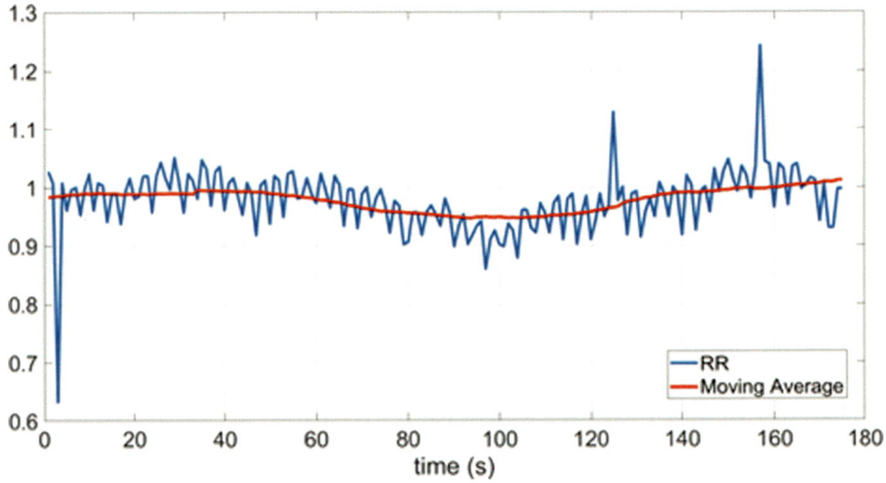

Fig. 5 R-R series over moving average process

this case, the time corresponding to the k+1th index sample position is also stored. If the above difference exceeds the threshold of 160 ms, the reference sample remains at the value of RR[k], discards RR sample [k+1] and continues scanning with the next sample in the RR series. By doing this, at the end, we have stored the valid values of the RR series and the corresponding time instants of the valid samples.

Finally, we apply the cubic spline interpolation technique to replace the gaps resulting from the ectopic beat elimination process. Figure 6 illustrates the original RR series (the same as in the previous figure) and the corrected series, which we will now denote NN series (series of intervals between sinus or normal beats).

3.4 *Heart Rate Variability Metrics*

From this point on, we started obtaining the Heart Rate Variability (HRV) metrics themselves. First, we determine the metrics in the frequency domain. An interpolated version of the NN series is calculated by considering equispaced samples at a uniform sampling frequency 2 Hz, also considering the cubic spline algorithm, then obtaining the power spectral density through the Yule-Walker method [14], considering a total of 64 coefficients and the frequency bands: VLF (0–0.04 Hz), LF (0.04–0.15) and HF (0.15–0.4). Figure 7 illustrates, for the same series example already discussed in this section, the power spectral density obtained through the Yule-Walker method:

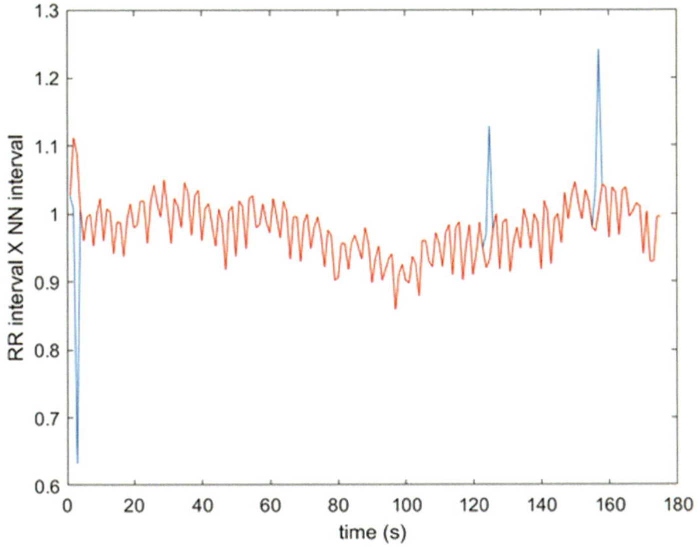

Fig. 6 R-R series and N-N series resulting from ectopic elimination processing

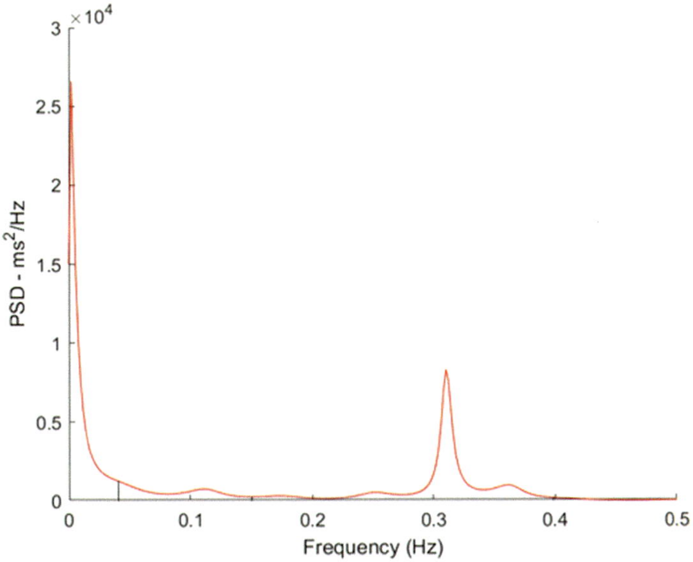

Fig. 7 Power spectral density, considering the VLF, LF and HF bands

Table 2 Description of the parameters (features) determined in the frequency domain

Parameter	Description
VLF Peak (Hz)	Location of the peak of the VLF component
LF Peak (Hz)	Location of the LF component peak
HF Peak (Hz)	Location of the HF component peak
VLF power (ms^2)	Area under the curve of the DEP function in the range corresponding to the VLF component
LF power (ms^2)	Area under the curve of the DEP function in the range corresponding to the LF component
HF power (ms^2)	Area under the curve of the DEP function in the band corresponding to the HF component
LF (nu)	Power of the LF component calculated in normalized units
HF (nu)	Power of the HF component calculated in normalized units
VLF (%)	Percentage of the area under the curve of the DEP function of the VLF component in relation to the total area
LF (%)	Percentage of the area under the curve of the DEP function of the LF component in relation to the total area
HF (%)	Percentage of the area under the curve of the DEP function of the HF component in relation to the total area
LF/HF(%)	Ratio between the absolute spectral powers of the LF and HF components

Frequency Domain Metrics

In this step, a number of parameters are determined: peaks of the VLF, LF and HF bands, areas of the individual spectral components, percentages relative to the total area (relative area), LF and HF components in normalized units and LF and HF components The relationship between [16]. We indicate in Table 2 the set of parameters (features) determined in the context of the frequency domain.

Time Domain Metrics

After computing the frequency domain parameters, we proceed to determine the metrics resulting from the time domain analysis, as shown in Table 3. It is important to emphasize that for the analysis in the time domain, we used a series of NN intervals (after correcting and filling the gap phase caused by discarding out-of-place beats). An interpolated version of the series (where the spacing between samples is uniform) is used for frequency domain analysis only.

Figure 8 illustrates the determination of the histogram of the series of NN intervals, the identification of the height of the histogram and the application of the necessary triangular interpolation to calculate the width of the base of the histogram, associated with the TINN variable.

Table 3 Description of metrics resulting from the time domain analysis

Parameter	Description
Average NN (s)	Average duration of NN intervals
SDNN (s)	Standard deviation of NN intervals
AHR (beat/min)	Average heart rate (heart rate)
SDHR (beat/min)	Standard deviation of heart rate values
SDANN (s)	Standard deviation of the averages of NN intervals in all 5-min segments of the entire recording
SDNN index (s)	Mean of the standard deviations of all NN intervals for all 5-min segments of the entire recording
RMSSD (s)	The square root of the mean of the sum of the squares of differences between adjacent NN intervals
NN50 count	Number of pairs of adjacent NN intervals differing by more than 50 ms in the entire recording
pNN50 (%)	Number of pairs of adjacent NN intervals differing by more than 50 ms in the entire recording
HRV	Triangular Index Total number of all NN intervals divided by the height of the histogram of all NN intervals measured on a discrete scale with bins of 7.8125 ms (1/128 s)
TINN (s)	Baseline width of the minimum square difference triangular interpolation of the highest peak of the histogram of all NN intervals

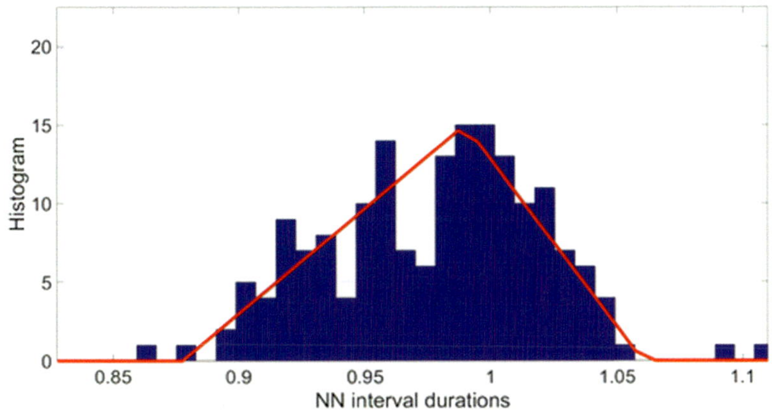

Fig. 8 Temporal analysis of Heart Rate Variability: histogram and obtaining the TINN metric

Non-linear Metrics

After determining the metrics in the time domain, we started to obtain the metrics in the context of non-linear analysis. It is known that [16] Nonlinear phenomena are certainly involved in the genesis of HRV. They are determined by complex interactions of hemodynamic, electrophysiological, and humoral variables as well as by the autonomic and central nervous regulations. It has been speculated that analysis of HRV based on the methods of nonlinear dynamics might elicit valuable information for physiological interpretation of HRV and for the assessment of the risk of sudden death. The parameters that have been used to measure nonlinear properties of HRV include 1/f scaling of Fourier spectra,47 19 H scaling exponent, and Coarse Graining Spectral Analysis (CGSA).48 For data representation, Poincaré sections, low-dimension attractor plots, singular value decomposition, and attractor trajectories have been used. For other quantitative descriptions, the D2 correlation dimension, Lyapunov exponents, and Kolmogorov entropy have been used.

First, we construct a Poincaré graph, which consists of a point cloud where each point is mapped to a location (x, y) = (NN[k], NN[k+1]), i.e. the current sample and the subsequent sample (= or predecessor) samples. We then calculated the standard deviation of the samples projected perpendicular to the identity line (SD1) and the identity line itself (SD2), respectively.

Figure 9 illustrates a Poincaré plot and a superimposed ellipse obtained for the same set of NN intervals discussed in this chapter, where the axes are proportional to the SD1 and SD2 measurements.

Next, sample entropy (**SampEn**) and approximate entropy (**ApEn**) are computed, which are widely used in the literature [17]. Both measure the regularity and complexity of the time series, where low values are often associated with the predictability of the signal. In addition to obtaining the relative entropy calculation of the original series, the calculation of entropies is also applied in versions of the signal derived from a process which we call multiscaling. Thus, let be the original series NN[k]. From it, we can get the series:

$$NN^{(1)}[k] = \frac{NN[2*k] + NN[2*k+1]}{2}$$

After calculating the ApEn and SampEn entropies for the series $NN^{(1)}[k]$, the series $NN^{(2)}[k]$ is obtained using the same recurrence relation. Thus, we have the generic derivation equation:

$$NN^{(j)}[k] = \frac{NN^{j-1}[2*k] + NN^{j-1}[2*k+1]}{2}$$

Preserving multiscale entropy has also been widely used in the literature [18]. For the time series in question (heart rate variability signal with a duration of five minutes), we obtain the ApEn and SampEn entropies, scalable up to level 5. In

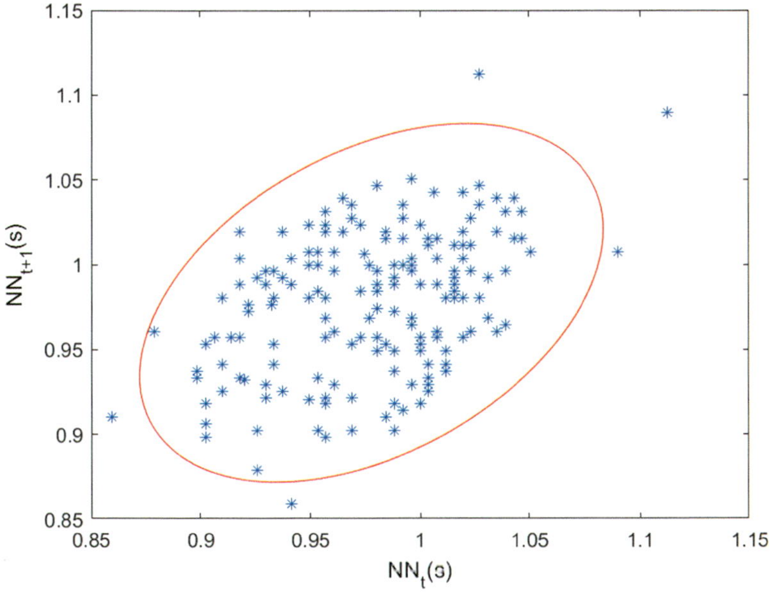

Fig. 9 Poincaré plot illustrating the determination of the SD1 (Short Term HRV) and SD2 (Long Term HRV) coefficients

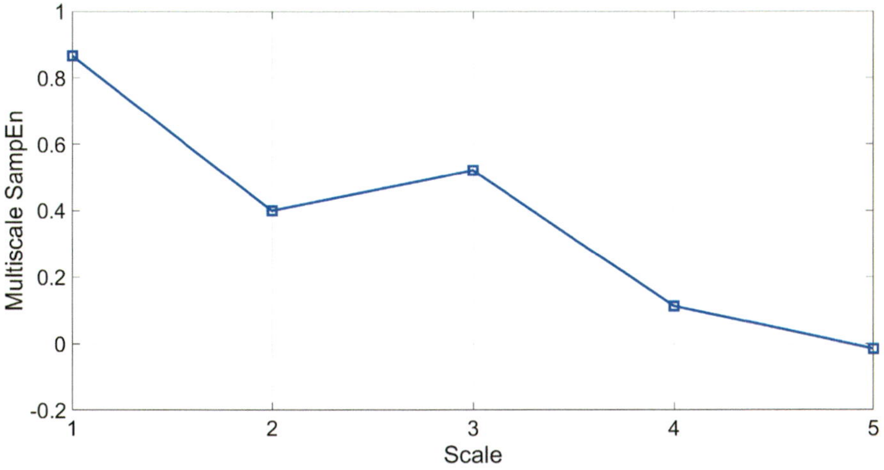

Fig. 10 Example of multiscale entropy calculation considering the SampEn entropy

Fig. 10, we obtain the time-based multiscale SampEn entropy calculation example series exemplified earlier.

Metrics derived from the PRSA (Phase Rectified Signal Averaging) method are calculated: PRSA DC and PRSA AC [19]. The fundamental principle of the PRSA

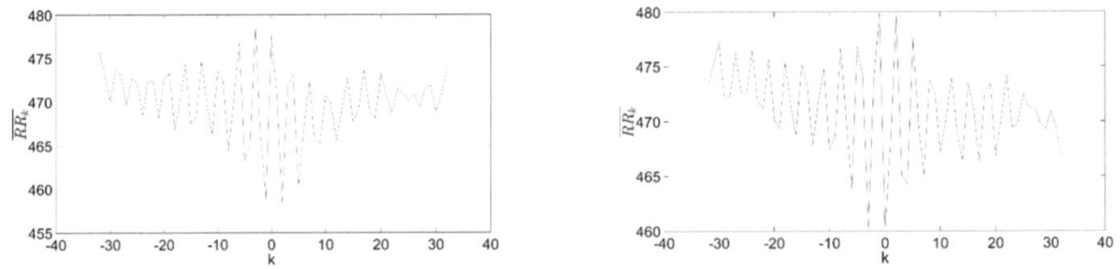

(a) Deceleration (b) Acceleration

Fig. 11 Results obtained for the application of the PRSA technique considering the anchor points

technique is to align segments of a time series with predefined features, which we call anchors. The method consists of three steps. In the first step, RR intervals are divided into two categories: the deceleration anchors (defined as intervals longer than the preceding intervals: Eq. 1) and acceleration anchors (defined as intervals shorter than the preceding intervals: Eq. 2).

$$RR[k] > RR[k-1] \tag{1}$$

$$RR[k] < RR[k-1] \tag{2}$$

In the second step, a window of size $2 * L$ (where L is the characteristic parameter of the method) is set around each anchor point, and the anchor point is located at the center of the window. The value of L must be greater than the period of the longest oscillation you wish to detect. Anchor points near the end of the signal that cannot generate a complete fragment of size $2 * L$ are ignored. If we consider the position of each anchor point as $\lambda_v, v = 1, ..., M$, the points forming the window v will be:

$$RR_{\lambda_v - L}, RR_{\lambda_v - L+1}, ..., RR_{\lambda_v + L-2}, RR_{\lambda_v + L-1}$$

In the third step, a PRSA signal is obtained for each set of anchor points (acceleration and deceleration) with the average of the rectified phase signal $\overline{(RR_k)}$, aligning all windows to the corresponding anchor points. Then, after stacking the windows, calculate the average value of the corresponding $RR[k]$ values to obtain the average window $\overline{(RR_k)}$, getting $PRSA - DC$ (deceleration) and $PRSA - AC$ (acceleration) coefficients as:

$$PRSA - DC(PRSA - AC) = \frac{1}{4}\overline{(RR_0)} + \overline{(RR_1)} - \overline{(RR_{-1})} - \overline{(RR_{-2})}$$

Figure 11a, b illustrate the application of the described technique.

On the last step we finally obtain the metrics **RR**, **DET** (determinism analysis coefficient in recurrence plot), **ENT**, **Lmax**, **L**, **TT** and **Vmax** [20, 21].

Table 4 Description of non-linear parameters. Adapted from [20]

Parameter	Description
ApEn	Measures the regularity and complexity of a time series
D_2	Estimates the minimun number of variables required to construct a model of system dynamics
DET	Recurrence plot analysis determinism
$DF\alpha 1$	Describe short-term fluctuations
$DF\alpha 2$	Describe long-term fluctuations
REC	Recurrence rate
SampEn	Measures the regularity and complexity of a time series
SD1	Measures ST HRV and is associated with baroreflex sensitivity (BRS)
SD2	Measures ST and LT HRV and is associated with LF absolute power and BRS
ShanEn	Measures the average information in a time series

3.5 Compilation of HRV Metrics

The application of the algorithms described in Sect. 3 gave us 43 metrics: Pico VLF, Pico LF, Pico HF, Power VLF, Power LF, Power HF, Power NU/LF, Power NU/HF, Power % VLF, Power % LF, Power % HF, LF/HF (%), Average NN (s), Stdev NN (s), Average HR (1/min), Stdev HR (1/min), SDANN (s), SDNN index (s), RMSSD (s), NN50 (n), pNN50 (%), Heartbeats (n), HRV index, TINN (s), Sample Entropy, Approximate Entropy, PRSA DC, PRSA AC, SD1 Poincare, SD2 Poincare, SampEn MSE 1, SampEn MSE 2, ApEn MSE 1, ApEn MSE 2, ApEn MSE 3, ApEn MSE 4, ApEn MSE 5, RR, DET, ENT, Lmax, L and Vmax. All these features was explained at Tables 2, 3, and 4.

4 Machine Learning Stage

4.1 Importance of Leave One Subject Out

The work done in [22] helped us understand the need to separate the training and test sets based on the subject rather than the exam. As presented in the article, this care is essential for the effectiveness of the classifier in real-life situations, and after verification tests involving adulteration of real data, it was proven that using the conventional k-fold makes the classifier learn about the subject and not on the facts related to the characteristics of the studied disease, strongly contaminating the final result. Results of applied comparisons showed that there was a significant drop in accuracy. In the case studied, the drop was from 98% to 70%. In addition, the results also showed a brutal difference in the standard deviation, obtaining better-behaved values for the k-fold, with deviations between ± 3 and ± 10, while for LOSO, the deviations reached values around ± 36. Based on this knowledge, we base our training strategy on this foundation, but as will be presented below, due to the low number of samples in our database, we allow ourselves to perform executions with a default leave one out. These differences in training will be explained in each case.

4.2 Dataset Information and Data Correlation

We already explained in Sect. 2 how the ECG data was collected and in Sect. 3 how it was processed. Table 5 presents our entire dataset information. The "PACIENTE" column contains the patient's name and its value are not mandatory unique since two ECG collections were performed for each one of them. This column help us to realize Leave One Subject Out style cross-validation. The attribute "gravidade" is considered the target for disease severity ranking.

Correlation explains how a variable's behavior is related to another's behavior. That is, it indicates if there is a relationship between two variables, for example, if changes in one variable will result in changes in the other one, in the same direction, or the opposite direction, considering a linear or non-linear behavior. The stronger the relationship, the more easily one variable will predict or influence the behavior of the other. Figure 12 shows us a heatmap of the correlations between dataset features. The darker the color, the stronger the correlation. Then our picture shows few variables that are strongly correlated. This result is expected and desirable for real-life data.

4.3 Training Strategies and Their Results

We selected 19 algorithms implemented in scikit-learn [15], a toolbox from the Python software development environment:

- BaggingClassifier
- GaussianProcessClassifier(1.0 * RBF(1.0))
- LinearSVC
- ExtraTreesClassifier($n_estimators = 300$)
- SGDClassifier($max_iter = 100$, $tol = 1e - 3$)
- SGDClassifier
- RandomForestClassifier
- DecisionTreeClassifier($max_depth = 5$)
- LogisticRegression(solver='lbfgs')
- SVC(gamma='auto')
- MLPClassifier(alpha=1, $max_iter = 1000$)
- GradientBoostingClassifier
- KNeighborsClassifier

Table 5 Dataset information containing data types for each feature

#	Column	Non-Null Count	Dtype
0	PACIENTE	98 non-null	Object
1	Pico VLF	98 non-null	Float64
2	Pico LF	98 non-null	Float64
3	Pico HF	98 non-null	Float64
4	Power VLF	98 non-null	Float64
5	Power LF	98 non-null	Float64
6	Power HF	98 non-null	Float64
7	Power NU/LF	98 non-null	Float64
8	Power NU/HF	98 non-null	Float64
9	Power % VLF	98 non-null	Float64
10	Power % LF	98 non-null	Float64
11	Power % HF	98 non-null	Float64
12	LF/HF	98 non-null	Float64
13	Average NN (s)	98 non-null	Float64
14	Stdev NN (s)	98 non-null	Float64
15	Average HR (1/min)	98 non-null	Float64
16	Stdev HR (1/min)	98 non-null	Float64
17	SDANN (s)	98 non-null	int64
18	SDNN index (s)	98 non-null	int64
19	RMSSD (s)	98 non-null	Float64
20	NN50 (n)	98 non-null	int64
21	pNN50 (%)	98 non-null	Float64
22	Heart beats (n)	98 non-null	int64
23	HRV index	98 non-null	Float64
24	TINN (s)	98 non-null	Float64
25	Sample Entropy	98 non-null	Float64
26	Approximate Entropy	98 non-null	Float64
27	PRSA DC	98 non-null	Float64
28	PRSA AC	98 non-null	Float64
29	SD1 Poincare	98 non-null	Float64
30	SD2 Poincare	98 non-null	Float64
31	SampEn MSE 1	98 non-null	Float64
32	SampEn MSE 2	98 non-null	Float64
33	ApEn MSE 1	98 non-null	Float64
34	ApEn MSE 2	98 non-null	Float64
35	ApEn MSE 3	98 non-null	Float64
36	ApEn MSE 4	98 non-null	Float64
37	ApEn MSE 5	98 non-null	Float64
38	RR	98 non-null	Float64
39	DET	98 non-null	Float64
40	ENT	98 non-null	Float64
41	Lmax	98 non-null	int64
42	L	98 non-null	Float64
43	Vmax	98 non-null	int64
44	Gravidade	98 non-null	int64

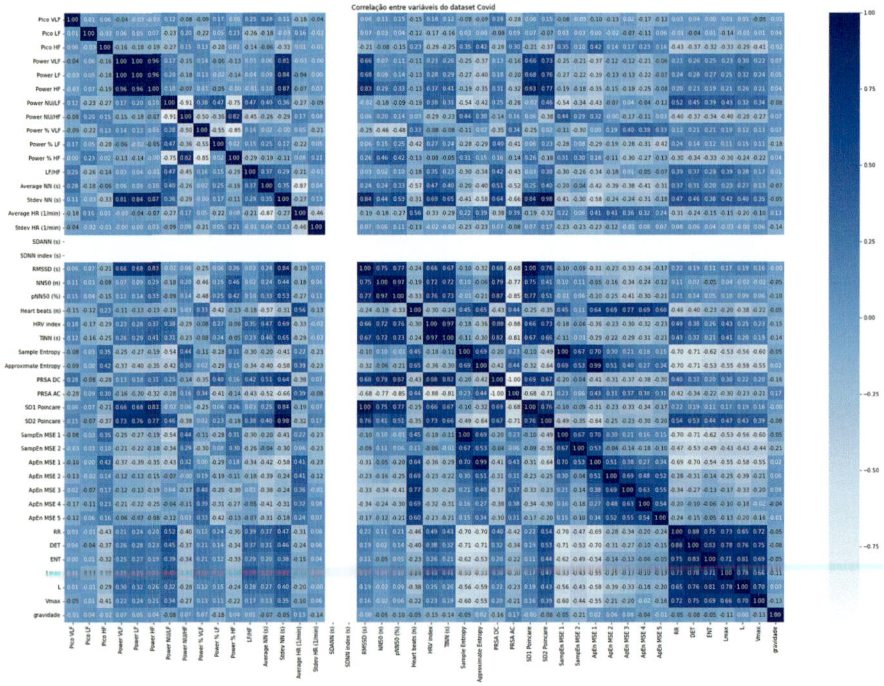

Fig. 12 Heatmap of the correlation between features extracted from the ECG dataset

- AdaBoostClassifier
- LGBMClassifier
- LinearDiscriminantAnalysis
- OneVsRestClassifier(LinearSVC($random_state = 0$, dual=False))
- QuadraticDiscriminantAnalysis
- GaussianNB.

We have developed a python tool(a object oriented class) that helps us apply different types of training and testing strategies for classification or regression, sampling, separation of training sets, testing and validation, based on subject for clinical cases or not. The source code (in Portuguese) is available online at https://github.com/pordeus/springer/. We describe each strategy used for this work in the following sections.

4.4 General Comparison

As a first evaluation, we performed a general comparison of the results of 19 algorithms, using three sizes of training and test sets, in the range of 70 and 30%, 80/20 and 90/10. For each training-test pair, we run all algorithms only once and compute the accuracy and the f1-score. It is important to emphasize that the separation of these sets for this evaluation respected the issue of learning about the disease and not about the patient (subject), that is, the exams of a patient were always on the same side, either from the training or the test. At the end of the process, the average for each algorithm was calculated and the result is presented in the Table 6 and they are sorted by accuracy. The configuration of the selected algorithms was applied according to the list presented in Sect. 4.3.

We can highlight in this step the results obtained by the algorithms LinearSVC, with higher accuracy and good value for F1-score, SVC and QuadraticDiscriminantAnalysis, with 67% for accuracy but incredible 1.00 for F1-score. In general we achieve good results for almost all model selected.

Table 6 Results obtained in general training

Model	Accuracy (%)	F1 Score
LinearSVC	71	0.76
LogisticRegression	70	0.82
MLPClassifier	70	0.70
SGDClassifier	69	0.80
SGDClassifier	69	0.80
OneVsRestClassifier	68	0.75
KNeighborsClassifier	68	0.85
QuadraticDiscriminantAnalysis	67	1.00
SVC	67	1.00
GaussianProcessClassifier	67	0.15
DecisionTreeClassifier	66	0.66
ExtraTreesClassifier	65	0.83
LGBMClassifier	64	0.74
RandomForestClassifier	64	0.83
BaggingClassifier	63	0.66
LinearDiscriminantAnalysis	61	0.70
GradientBoostingClassifier	61	0.72
AdaBoostClassifier	60	0.71
GaussianNB	20	0.08

Table 7 Results for leave one subject out training

Model	Accuracy (%)	F1 Score
QuadraticDiscriminantAnalysis	66	0.79
SVC	66	0.79
LinearDiscriminantAnalysis	63	0.70
ExtraTreesClassifier	63	0.74
LogisticRegression	63	0.74
RandomForestClassifier	62	0.75
OneVsRestClassifier	61	0.72
KNeighborsClassifier	61	0.71
SGDClassifier	60	0.69
MLPClassifier	59	0.66
SGDClassifier	58	0.66
LinearSVC	56	0.59
GradientBoostingClassifier	55	0.65
LGBMClassifier	54	0.64
BaggingClassifier	52	0.62
DecisionTreeClassifier	50	0.61
AdaBoostClassifier	47	0.59
GaussianProcessClassifier	37	0.20
GaussianNB	33	0.13

4.5 *Leave One Subject Out*

Each round performed training with 49 patients, leaving one out for validation. At the end of all checks, the average of accuracy and f1-score is calculated for each algorithm. Table 7 brings the results obtained.

4.6 *K-Fold Training*

Because the dataset (after cleaning) has 98 exam samples from 50 patients. We implemented a training using K-Fold using the factorization of 98 between 7 and 14. We processed the training using two windows, one with 7 partitions of 14 samples each, and another with 14 partitions of 7 samples. All algorithms are executed for the two windows and averaged accuracy and f1-score are calculated. The results are presented in Table 8. In this training, the verifications of the rule per patient were not followed. The Fold window counted the number of exams.

We can highlight the results of SVC and QuadraticDiscriminantAnalysis, but in addition, we draw attention to results, in general, of accuracy values close to those

Table 8 Results for K-Fold Training

Model	Accuracy (%)	F1 Score
SVC	66	0.78
QuadraticDiscriminantAnalysis	66	0.79
ExtraTreesClassifier	66	0.76
LogisticRegression	63	0.74
KNeighborsClassifier	62	0.72
LinearDiscriminantAnalysis	62	0.69
OneVsRestClassifier	62	0.73
LinearSVC	61	0.65
RandomForestClassifier	61	0.74
SGDClassifier	57	0.65
SGDClassifier	56	0.66
BaggingClassifier	56	0.66
MLPClassifier	55	0.64
AdaBoostClassifier	52	0.65
LGBMClassifier	52	0.63
GradientBoostingClassifier	51	0.64
DecisionTreeClassifier	50	0.57
GaussianProcessClassifier	35	0.16
GaussianNB	29	0.11

of general training, which considers the issue of the Subject. Of course, they were very different approaches, but it did not appear to us that there was contamination of the test. Probably, due to the fact that there are a maximum of two exams per patient and that the exams are contiguous, there must have been few cases where the same patient had the exams separated in training and testing.

5 Discussion

As demonstrated in Sect. 4, the effectiveness of classifying the severity of COVID 10 by applying machine learning algorithms on linear and non-linear metrics extracted from electrocardiogram (ECG) exams showed accuracy rates between 65 and 70% in different training strategies, always having at least five algorithms reaching F1-Score values above 0.70, with a peak of 1.00. Considering that the dataset used for the analyzes was small, with a total of 50 volunteers, 98 samples, and 43 features (in addition to being unbalanced between the different classes), the proposed system achieved promising results, achieving the best performance with the Support Vector Machine Classifier.

Even having performed one of the training rounds without separating by subject, due to the low number of samples available, we believe the contamination result was minimal. Still, as we have an increase in the database, the differences in the results will possibly become more visible, confirming the need for care with learning focused on the disease and reinforcing confidence in the results that will be obtained.

6 Conclusion

A tool for binary classification according to the degree of severity of hospitalized patients diagnosed with Covid-19 seems plausible. The groups is selected with patients classified as mild and moderate, while the second group is for patients classified as severe or critical. Determining two groups became necessary due to the limited number of samples and their imbalance. The proposed solution achieved an accuracy of 71% and F1-score of 1.00.

We were cautious with the preparation and training of our research, given the subject's sensitivity, which gives us confidence in the results obtained and leaves room for the search for improvements.

In general, we highlight the findings for the SVC algorithm. The strategy to continue this work will be to focus on increasing the performance of the algorithms used in this study. Since there are many indices (features) of HRV in relation to the sample size [23], some of the features of HRV may not be relevant to our classification problem. With that in mind, the best features can be searched for the best results presented in this work, using the feature selection scheme and grid search.

Acknowledgements Joao Paulo Madeiro would like to thank Funcap (Ceara State Foundation for the Support of Scientific and Technological Development) Grant PS1-0186-00439.01.00/21.
Joao Alexandre Lobo Marques would like to thank the research team from the Laboratory of Applied Neurosciences at the University of Saint Joseph, Macau SAR, China.

References

1. Zheng YY, Ma YT, Zhang JY, Xie X (2020) COVID-19 and the cardiovascular system. Nat Rev Cardiol 17:259–260
2. Hosseini E et al (2020) The novel coronavirus disease-2019(COVID-19): mechanism of action, detection and recent therapeutic strategies. Virology 551:1–9
3. Lai C, Lam W (2021) Laboratory testing for the diagnosis of COVID-19. Biochem Biophys Res Commun 538:226–230
4. Wang S et al (2020). A fully automatic deep learning system for COVID-19 diagnostic and prognostic analysis. Eur Respirat J 56(2):2000775. https://doi.org/10.1183/13993003.00775-2020
5. Khuzani Z et al (2021) COVID-Classifier: an automated machine learning model to assist in the diagnosis of COVID-19 infection in chest X-ray images. Sci Rep J 11(6). https://doi.org/10.1038/s41598-021-88807-2

6. Shaffer F, McCraty R, Zerr CL (2014) A healthy heart is not a metronome: an integrative review of the heart's anatomy and heart rate variability. Front Psychol 5:1040. https://doi.org/10.3389/fpsyg.2014.01040
7. Mccraty R, Shaffer F (2015) Heart rate variability: new perspectives on physiological mechanisms, assessment of self-regulatory capacity, and health risk. Global Adv Health Med 4(1):46–61. https://doi.org/10.7453/gahmj.2014.073
8. YAN X et al (2020) Clinical characteristics and prognosis of 218 patients with COVID-19: a retrospective study based on clinical classification. Front Medi 7:485
9. Felber Dietrich D et al (2006) Heart rate variability in an ageing population and its association with lifestyle and cardiovascular risk factors: results of the SAPALDIA study. Europace 8(7):521–529
10. Sj B, Ja B (2007) Resting and postexercise cardiac autonomic control in trained masters athletes. J Physiol Sci 57(1):23–29
11. Vonesch C; Blu T, Unser M (2007) Generalized Daubechies wavelet families. IEEE Trans Signal Process 55(9):4415–4429
12. Madeiro JPV (2012) An innovative approach of QRS segmentation based on first-derivative, Hilbert and wavelet transforms. Med Eng Phys 34(9):1236–1246
13. Hou T, Qin H (2012) Continuous and discrete Mexican hat wavelet transforms on manifolds. Graph Models 74(4):221–232
14. Berntson Gary G et al (1997) Heart rate variability: origins, methods, and interpretive caveats. Psychophysiology 34(6):623–648
15. Pedregosa F, Varoquaux G, Gramfort A, Michel V, Thirion B, Grisel O, Blondel M, Prettenhofer P, Weiss R, Dubourg V, Vanderplas J, Passos A, Cournapeau D, Brucher M, Perrot M, Duchesnay E (2011) Scikit-learn: machine learning in python. J Mach Learn Res V 12:2825–2830
16. Heart rate variability: standards of measurement, physiological interpretation and clinical use. Task Force of the European Society of Cardiology and the North American Society of Pacing and Electrophysiology. Circulation. 1996 Mar 1;93(5):1043-65. PMID: 8598068
17. Richman JS, Moorman JR (2000) Physiological time-series analysis using approximate entropy and sample entropy. Am J Physiol-Heart and Circulat Physiol 278(6):H2039–H2049
18. Costa M, Goldberger AL, Peng C-K (2005) Multiscale entropy analysis of biological signals. Phys Rev E 71(2):021906
19. Schneider R, Müller A, Schmidt G (2012) Phase-rectified signal averaging: methods and clinical applications. Heart Rate Variability (HRV) Signal analysis: clinical applications, 1st edn. CRC, Boca Raton, FL, pp 28–61
20. Shaffer F, Meehan ZM, Zerr CL (2020) A critical review of ultra-short-term heart rate variability norms research. Front Neurosci 14:594880
21. Shinoda L et al (2021) Cardiac and autonomic dysfunctions assessed through recurrence quantitative analysis of electrocardiogram signals and an application to the 6-hydroxydopamine Parkinson's disease animal model. Front Physiol 1919p
22. Kunjan S et al (2021) The necessity of leave one subject out (LOSO) cross validation for EEG disease diagnosis. In: Mahmud M, Kaiser MS, Vassanelli S, Dai Q, Zhong N (eds) Brain informatics. BI 2021. Lecture notes in computer science, vol 12960. Springer, Cham
23. Silva LEV, Moreira HT, de Oliveira MM, Cintra LSS, Salgado HC, Fazan R Jr, Tinós R, Rassi A Jr, Schmidt A, Marin-Neto JA (2022) Heart rate variability as a biomarker in patients with Chronic Chagas Cardiomyopathy with or without concomitant digestive involvement and its relationship with the Rassi score. Biomed Eng Online. 21(1):44. https://doi.org/10.1186/s12938-022-01014-6.PMID: 35765063; PMCID: PMC9241264. Accessed from 28 June 2022

Exploratory Data Analysis on Clinical and Emotional Parameters of Pregnant Women with COVID-19 Symptoms

Joao Alexandre Lobo Marques, Danielle S. Macedo, Pedro Motta, Bruno Riccelli dos Santos Silva, Francisco Herlanio Costa Carvalho, Renata Castro Kehdi, Letícia Régia Lima Cavalcante, Marylane da Silva Viana, Deniele Lós, and Natália Gindri Fiorenza

Abstract The scientific literature indicates that pregnant women with COVID-19 are at an increased risk for developing more severe illness conditions when compared with non-pregnant women. The risk of admission to an ICU (Intensive Care

---

J. A. Lobo Marques (✉)
Laboratory of Applied Neurosciences, University of Saint Joseph, West Hartford, CT, USA
e-mail: alexandre.lobo@usj.edu.mo

D. S. Macedo
Department of Physiology and Pharmacology, Faculty of Medicine,
Federal University of Ceará, Fortaleza, Brazil
e-mail: danielle.macedo@ufc.br

P. Motta · B. R. dos Santos Silva
LESC, Federal University of Ceará, Fortaleza, Brazil
e-mail: pedro.motta@lesc.ufc.br

B. R. dos Santos Silva
e-mail: bruno@lesc.ufc.br

F. H. C. Carvalho
Assis Chateaubriand Maternity Public Hospital (MEAC), Faculty of Medicine,
Federal University of Ceará, Fortaleza, Brazil
e-mail: herlaniocosta@ufc.br

R. C. Kehdi · L. R. L. Cavalcante · M. da Silva Viana
Faculty of Medicine, Federal University of Ceará, Fortaleza, Brazil
e-mail: renatakehdi@hotmail.com

L. R. L. Cavalcante
e-mail: leticiaregia@outlook.com

M. da Silva Viana
e-mail: marylanedasilvaviana@gmail.com

D. Lós · N. G. Fiorenza
Núcleo de Pesquisa e Desenvolvimento de Medicamentos, Federal University of Ceará,
Fortaleza, Brazil
e-mail: deniele.los@gmail.com

N. G. Fiorenza
e-mail: nataliagfiorenza@gmail.com

J. A. Lobo Marques and S. J. Fong (eds.), *Computerized Systems for Diagnosis and Treatment of COVID-19*, https://doi.org/10.1007/978-3-031-30788-1_11

Unit) and the need for mechanical ventilator support is three times higher. More significantly, statistics indicate that these patients are also at 70% increased risk of evolving to severe states or even death. In addition, other previous illnesses and age greater than 35 years old increase the risk for the mother and the fetus, including a higher number of cesarean sections, higher systolic and diastolic maternal blood pressure, increasing the risk of eclampsia, and, in some cases, preterm birth. Additionally, pregnant women have more Emotional lability/fluctuations (between positive and negative feelings) during the entire pregnancy. The emotional instability and brain fog that takes place during gestation may open vulnerability for neuropsychiatric symptoms of long COVID, which this population was not studied in depth. The present Chapter characterizes the database presented in this work with clinical and survey data collected about emotions and feelings using the Coronavirus Perinatal Experiences—Impact Survey (COPE-IS). Pregnant women with or without COVID-19 symptoms who gave birth at the Assis Chateaubriand Maternity Hospital (MEAC), a public maternity of the Federal University of Ceara, Brazil, were recruited. In total, 72 mother-infant dyads were included in the study and are considered in this exploratory analysis. The participants have undergone serological tests for SARS-CoV-2 antibody detection and a nasopharyngeal swab test for COVID-19 diagnoses by RT-PCR. A comprehensive Exploratory Data Analysis (EDA) is performed using frequency distribution analysis of multiple types of variables generated from numerical data, multiple-choice, categorized, and Likert-scale questions.

1 Introduction

The medical and scientific literature indicates that pregnant patients with COVID-19 are at an increased risk for developing more severe illness conditions when compared with nonpregnant patients. The risk of admission to an ICU (Intensive Care Unit) and the need for mechanical ventilator support is three times higher. More significantly, statistics indicate that these patients are also at 70% increased risk of evolving to severe states or even death. In addition, other previous illnesses and age greater than 35 years old increase the risk for the mother and the fetus, including a higher number of cesarean sections, higher systolic and diastolic maternal blood pressure, increasing the risk of eclampsia, and, in some cases, preterm birth [1].

The National Health System (NHS) identified a set of risk factors [2], according to a vast data collection in the UK and worldwide, to raise the attention of patients under any of the following conditions:

- more than 28 weeks pregnant (third trimester);
- Diabetes;
- Asthma;
- High blood pressure (Hypertension);
- Heart disease;
- Overweight;

- more than 35 years old;
- belongs to an ethnic minority group.

Several initiatives are being developed by societies, maternity hospitals, and clinics to support this group of patients. The American College of Obstetricians and Gynecologists (ACOG) and the Society of Maternal-Fetal Medicine prepared a manual of recommendations for medical teams working with pregnant women who might have COVID-19 or have been diagnosed with the illness. These guidelines are under constant revision and updated to reflect the most recent advances and updates regarding the disease characteristics, infection waves, new variants of the virus, or even new public health measures [1].

Another important initiative was the COVID generation (COVGEN) Research Alliance, which focuses on identifying the consequences of the pandemic for young families and children conceived, carried, and born during the pandemic [3].

The long-term effects of gestational exposure to Sars-CoV-2 are not yet fully understood as well as seeks solutions to mitigate negative effects. However, the effects of immune system activation during pregnancy are already well-known in preclinical and clinical studies, related to an increased risk of developing neurodevelopmental disorders such as autism spectrum disorder and schizophrenia [4, 5].

The main objective of this Chapter is to characterize the database presented in this work with clinical and survey data on emotions and feelings collected from pregnant women admitted to the Assis Chateaubriand Maternity Hospital (MEAC) MEAC Maternity Hospital. A comprehensive Exploratory Data Analysis (EDA) is performed using frequency distribution analysis of multiple types of variables, generated from numerical data, multiple-choice, categorized, and Likert-scale questions.

2 Related Works

Most studies addressing the consequences of SARS-CoV-2 infection for the fetus or neonate are case reports or case series, as expected for a pandemic that has been recently active. Most studies reinforce the importance of medium and long-term studies to understand the possible effects on the fetus of maternal gestational involvement by SARS-CoV-2 [6].

Studies indicate that, although apparently SARS-CoV-2 is rarely transmitted transplacentally to the fetus, infection with the COVID-19 virus during pregnancy may be linked to several adverse pregnancy outcomes. A systematic review and meta-analysis conducted in Canada, considering only publications with appropriate comparison groups, found a higher risk of preeclampsia, preterm birth, and stillbirth among pregnant women with SARS-CoV-2 infection than those without SARS-CoV-2 infection. In addition, among pregnant women infected with COVID-19, severe cases were more associated with severe outcomes and gestational issues, such as preeclampsia, preterm birth, gestational diabetes, and low birth weight compared with mothers suffering from mild symptoms [7].

From the social-economic perspective, normal epidemiological behavior is that the infection with SARS-CoV-2 is more likely to affect individuals living in socially and economically challenging conditions. Emeruwa et al. presented a report characterizing regions in New York City, indicating that infection rates were higher in pregnant women who lived in buildings with lower mean assessed values. It was also higher in the neighborhoods with lower median household incomes, higher unemployment rates, large household sizes, and greater household crowding [8]. In another document, Joseph et al. present that higher rates of COVID-19 infection among pregnant women were associated with Hispanic ethnicity, lack of health insurance, high neighborhood density, and paradoxically, smaller household size [9].

The use of medical imaging, together with clinical data, may indicate valuable information for pregnant women and newborns. In [10], the authors investigate clinical data, and Chest CT-Scan features in 59 patients, including 14 non-pregnant women, 41 pregnant women, and four children diagnosed with COVID-19. The findings indicate statistically significant differences between non-pregnant and pregnant groups in clinical parameters, such as leukocytosis, elevated neutrophil ratio, and lymphopenia. For image parameters, ground-glass opacity (GGO) was the predominant presence in 72% lesions for the non-pregnant group. Mixed and complete consolidation were more common on average (46%) in the pregnant group. Also, for the children group, effects detected in lung CT Scan were mild presenting focal GGO or consolidation, indicating that clinical analysis is possibly the most adequate protocol to assess COVID-19 in children.

Another important aspect to consider is the immunological response, its impact on maternal and fetal complications, and the evolution of severe prognosis during a COVID-19 infection. A comprehensive study in China by Hui et al. [11] indicates different risks for the fetus and the pregnancy term due to the maternal immunological response to the SARS-CoV-2 virus.

A comprehensive work following 1.416 cases of pregnant women, with 311, 433, and 672 patients in the first, second, and third trimesters of pregnancy, respectively, diagnosed with COVID-19 in Ankara in 2020–21 is presented in [12]. A total of 1.8% of the patients were admitted to the intensive care unit (ICU) and complications were present in around 16.1% of the cases, with preterm labor in 2.9%. Severe symptoms of the disease were found in pregnant women with no previous health problems. Mild symptoms of COVID-19 were more common in the first trimester while severe symptoms were more frequent in the second trimester. In addition, higher hospitalization rates were more common in the third trimester.

Finally, it should be mentioned that pregnant women, regardless of any comorbidity or infection, present emotional instability, as well as impairment in general cognitive, executive, and memory functioning, especially during the third trimester of pregnancy, the so-called "pregnancy brain fog" that seems to persist for a few months after delivery. If persistent, this condition may cause important consequences for the baby's development [13].

3 Materials and Methods

Since this work is focused on the exploratory analysis of data collected by a research team from the Neuropsychopharmacology laboratory of the Federal University of Ceara during the COVID-19 outbreak in 2021 and 2022, this Section will present the data source, describing relevant aspects of the database and the importance and characteristics of collected attributes.

3.1 Study Design

Pregnant women with or without covid-19 symptoms who gave birth at the Assis Chateaubriand Maternity Public Hospital at the Federal University of Ceará (MEAC-UFC), Brazil, were recruited. In total, 72 mother-infant dyads were included in the study and are considered in this exploratory analysis. The participants have undergone serological tests for SARS-CoV-2 antibody detection and a nasopharyngeal swab test for COVID-19 diagnoses by RT-PCR.

3.2 Patients' Selection

The inclusion criteria considered pregnant women admitted to the Emergency Service of the Assis Chateaubriand Maternity School, a public hospital from the Federal University of Ceará (MEAC-UFC), from January 2021 to August 2022. The patients 18 years old or over presented or did not have COVID-19 symptoms. The participants have undergone nasopharyngeal swab tests for COVID-19 diagnoses by RT-PCR and serological tests for SARS-CoV-2 antibody detection. The "control" group included asymptomatic women with no evidence of SARS-CoV-2 infection during pregnancy.

A flowchart with the process from entering the Emergency Room to the application of the Coronavirus Perinatal Experiences—Impact Survey (COPE-IS) is presented in Fig. 1.

The "symptomatic PCR+" group comprised pregnant women showing COVID-19 symptoms in the third trimester of pregnancy and a positive RT-PCR for SARS-CoV-2. The "symptomatic PCR-" group included patients with respiratory symptoms in the third trimester of pregnancy suggestive of COVID-19 infection but with a negative RT-PCR for SARS-CoV-2. The participants were submitted to the Portuguese version of the COPE-IS. According to the MEAC-UFC protocol, the acute respiratory condition was diagnosed by at least two of the following symptoms: fever, chills, sore throat, headache, cough, runny nose, smell disturbances, or taste disturbances. The worsening of the flu-like syndrome was considered when the patient additionally presented at least one of the following symptoms: dysp-

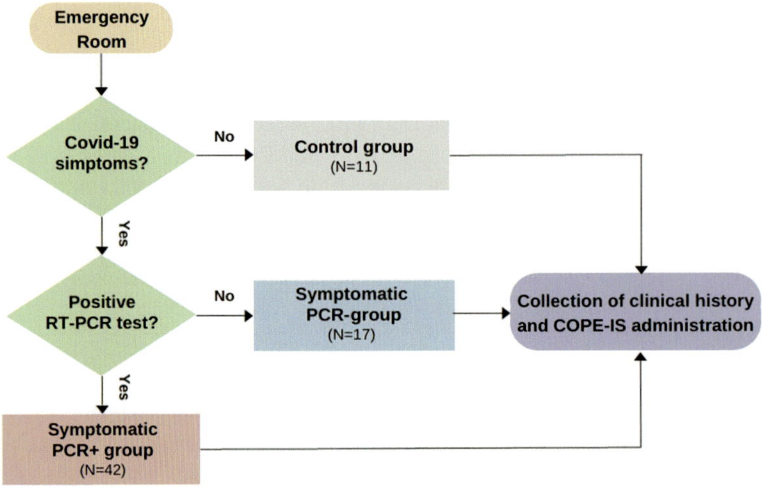

Fig. 1 Flow at the Maternity Hospital MEAC, from entering the Emergency Room to the application of the COPE Survey

nea/respiratory discomfort, persistent pressure in the chest, O_2 saturation lower than 95%, or bluish coloration of the lips or face.

3.3 COPE Survey

The group "All of Us Research Hub" developed a COPE-IS to test the impact of the COVID-19 pandemic on pregnancy outcomes and long-term maternal and child health. The questionnaire is composed of 53 questions, divided into nine sections. The first section of the questionnaire is different for pregnant and postpartum women. The other sections are similar for pregnant and postpartum women and comprise questions related to:

1. Perinatal experiences related to the COVID-19 outbreak;
2. COVID-19 exposures and symptoms (self and family);
3. COVID-19 financial considerations (current and expected future);
4. COVID-19 social support activities;
5. COVID-19 outbreak—restricted activities;
6. COVID-19 outbreak—coping and adjustment;
7. COVID-19 outbreak—emotions and feelings;
8. COVID-19 outbreak—health background, mental health, and substance use;
9. demographic background.

For this book chapter, we analyzed the answers to Sect. 7, with a focus on the psychological impacts of the pandemic on the participants. Depending on the question, responses may be given in a Yes/No format, selecting response items from those presented or in a Likert scale format. This instrument is available on an open science platform (https://osf.io/uqhcv/) with no restrictions on its rights or use.

3.4 *Groups of Study*

The subjects were divided into three groups according to the COVID-19 diagnostics throughout pregnancy. The "Control" group included women with no evidence of SARS-CoV-2 infection during pregnancy and who had received at least one dose of the vaccine. The "Symptomatic PCR+" group comprised pregnant women infected with SARS-CoV-2 in the third trimester of pregnancy (symptomatic pregnant and positive RT-PCR for SARS-CoV-2); and "Symptomatic PCR-" group included patients with presenting COVID-19 symptoms negative PCR result (symptomatic pregnant and negative RT-PCR for SARS-CoV-2).

3.5 *Data and Sample Collection Procedures*

As a first phase, data on the patient's obstetric history, health history, and socioeconomic status was collected upon enrollment in the study. Maternal blood collection was performed within 24 h of delivery. The Maternity Hospital nursing and neonatology team collected the umbilical cord blood at birth.

3.6 *Ethical Considerations*

This study followed the guidelines and regulatory standards for research involving human subjects and was approved by the Research Committee of the Assis Chateaubriand Maternity School, process number 4.492.367. Informed written consent was obtained from all participants or guardians for data and biospecimen collection. The study did not pose risks to the participants, except for the inconvenience of answering questions and collecting blood, as there was no intervention. The participants' autonomy was respected, as it was possible to withdraw at any time during the research.

4 Exploratory Data Analysis—EDA

As previously explained, this Chapter presents an Exploratory Data Analysis (EDA) of the three main groups listed as follows:

- "Control"—group included women with no evidence of SARS-CoV-2;
- "Symptomatic PCR+"—group comprised symptomatic pregnant and positive RT-PCR for SARS-CoV-2;
- "Symptomatic PCR-"—group included symptomatic pregnant and negative RT-PCR for SARS-CoV-2.

First, a sample characterization is performed, mostly using graphical analysis. Second, the symptoms are grouped and analyzed. After that, two essential dimensions for the analysis are presented, focusing on Maternal Health Status and Child Birth Parameters. In addition, following item 7 of the COPE-IS Questionnaire, maternal emotions and feelings are assessed. Finally, Correlation Matrixes between all variables and subgroups are presented to make it possible to confirm expected correlations and determine new correlations found in this database.

4.1 Dataset Characterization

In Fig. 2 the percentage of each group of participants is presented, with the Control group consisting of 15% of the total number of participants, while the Symptomatic groups are 85% of the total, with 61% Symptomatic with positive PCR results and 24% Symptomatic with negative PCR results.

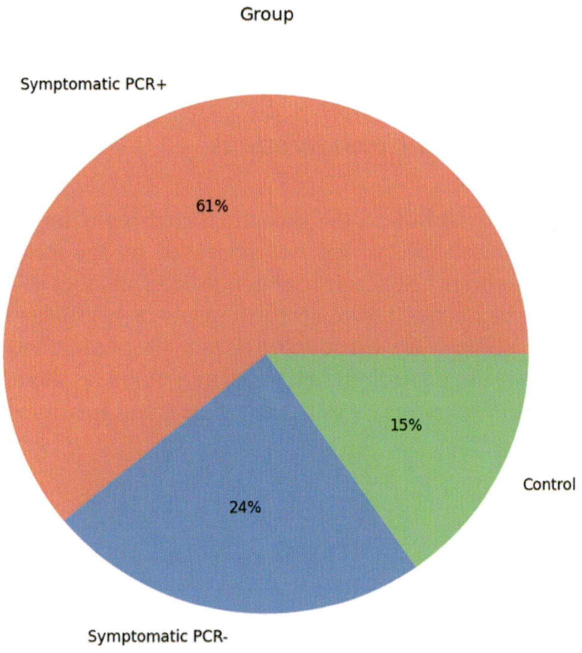

Fig. 2 Dataset characterization—groups of participants

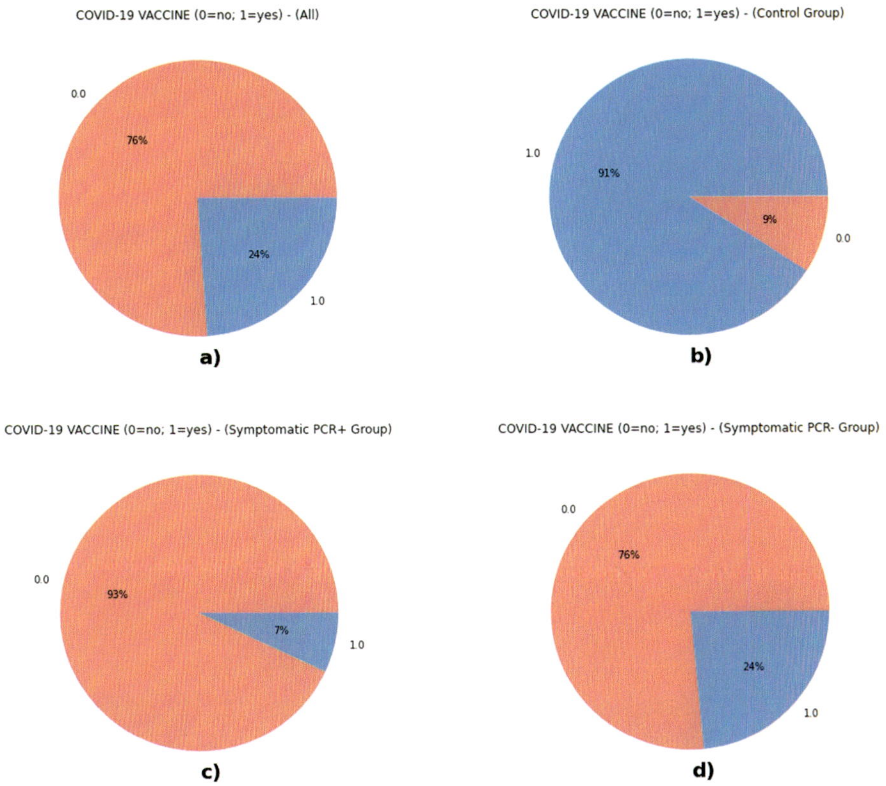

Fig. 3 Groups characterization—vaccine immunization. Frequency distribution analysis considering **a** All the participants, **b** Control group, **c** Symptomatic PCR+ group, and **d** Symptomatic PCR- group

In Fig. 3 the groups are presented according to the immunization with COVID-19 vaccines. It is important to highlight that these data were collected from January 2021 to August 2022 and those pregnant and lactating women were included in Brazil's priority group for vaccination against COVID-19 in May 2021. Considering the three groups together, the overall vaccine immunization rate was 24%. The Symptomatic PCR+ group only presented 9% of immunization. On the other hand, the Control group had 91% of immunization and the Symptomatic PCR- had 24%.

It is also relevant to mention that our COVID groups were predominantly collected until July 2021, when cases of the pandemic were increasing in Brazil, while the control group was collected predominately from July 2021 to August 2022. This may explain the higher number of vaccinated women in the control group.

Considering the groups' ethnicity, presented in Fig. 4, the frequency distribution was approximately similar for all the groups, with the Brown category representing 72% of the overall participants. It also accounted for the majority of the three groups. The second-largest ethnic group is White, with 21%, followed by Black, with 6%,

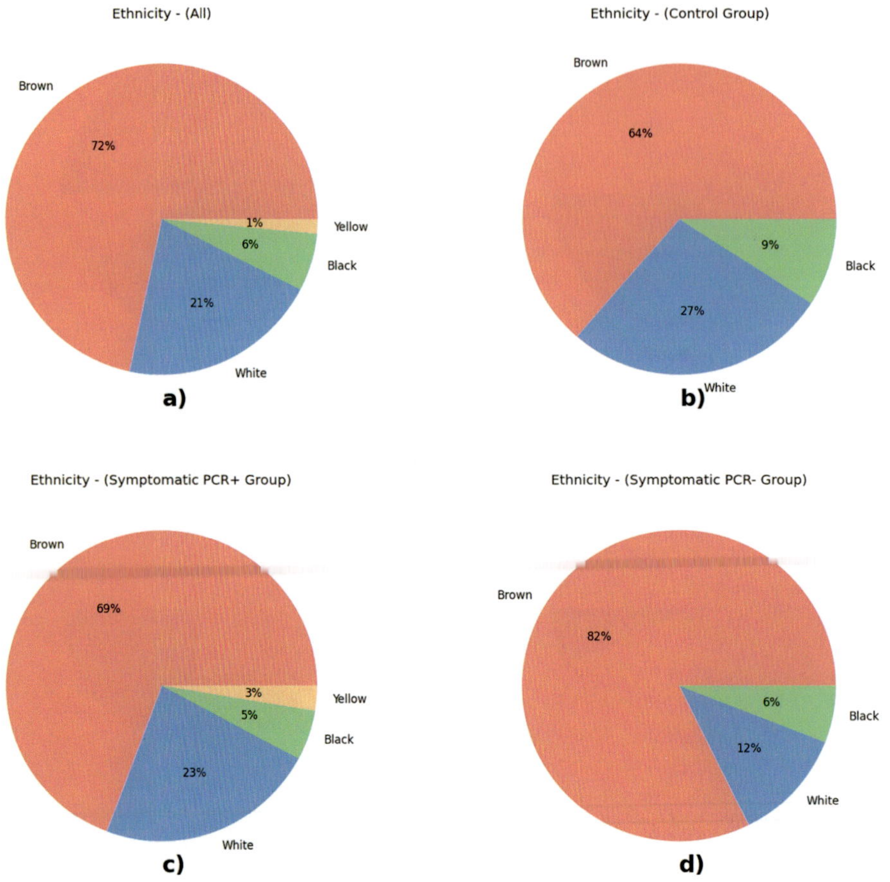

Fig. 4 Groups characterization—ethnicity. Frequency distribution analysis considering **a** All the participants, **b** Control group, **c** Symptomatic PCR+ group, and **d** Symptomatic PCR- group

and Yellow, with 1%. Detailed group information can be found in subfigures (b), (c), and (d).

Marital status characterization is presented in Fig. 5. In this study, Married and "Stable Union" were the most relevant categories, with a total percentage of 34% and 42%, respectively. The Control group only presented these two categories with 40% Married and 60% "Stable Union". Divorced was almost nonexistent, with only 2% of the total, and all were part of the Symptomatic PCR+ group, while Single participants were 23% of the total.

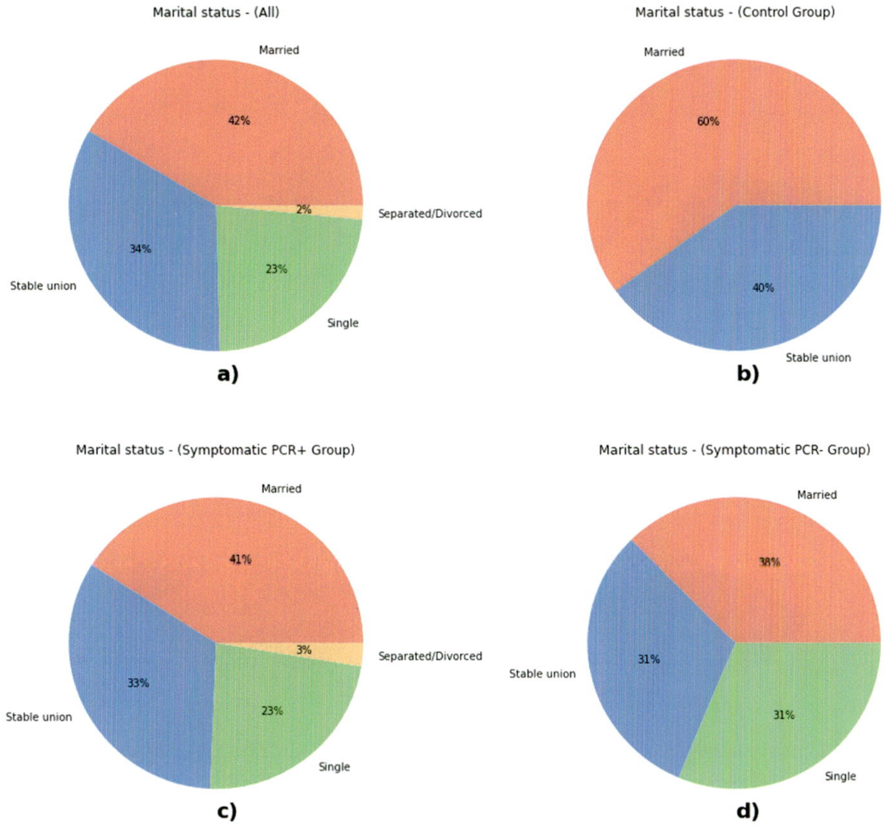

Fig. 5 Groups characterization—marital status. Frequency distribution analysis considering **a** all the participants, **b** Control group, **c** Symptomatic PCR+ group, and **d** Symptomatic PCR- group

Considering the educational stage of the participants may allow additional future subcategorization since it is usually correlated with the number of deliveries and access to better health support. This category is presented in Fig. 6. In this study, "Complete High School" and "Incomplete High School" were the most present categories, with a total percentage of 52% and 16%, respectively. The overall number for "Complete Higher Education" was 6%. In contrast, the Control group had 18% of participants with "Complete Higher Education", which may indicate a correlation with the higher number of Vaccinated participants in this group. The "Complete High School" was 58% of the Symptomatic PCR+ group, while the category of "Complete Primary Education" represented 20% of the Symptomatic PCR- group.

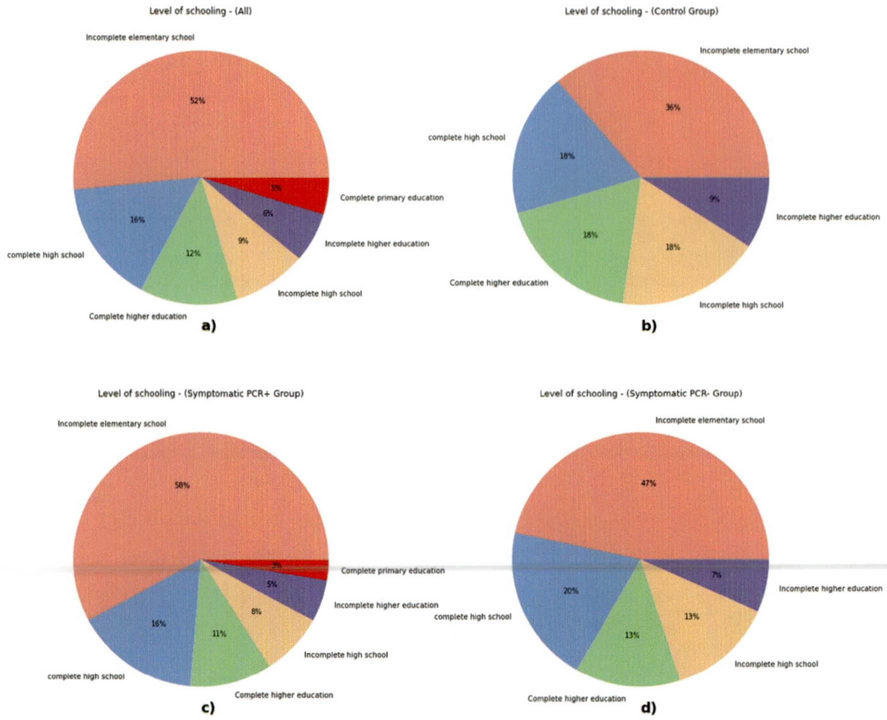

Fig. 6 Groups characterization—level of Schooling. Frequency distribution analysis considering **a** All the participants, **b** Control group, **c** Symptomatic PCR+ group, and **d** Symptomatic PCR- group

In the following two sets of plots, Figs. 7 and 8, the frequency analysis of the number of prior pregnancies and the number of deliveries are presented. A correlation between the level of schooling and marital status with these parameters may be expected. Considering the number of pregnancies, the highest frequency for the group with all participants was 2, followed by 3 and 1. Similar behavior is found for the Symptomatic PCR+ group. For the Symptomatic PCR- group, the discrepancy between the most frequent number of pregnancies, 2, was higher and was followed by 1 and 3. Finally, the Control group presented three as the most frequent number of pregnancies.

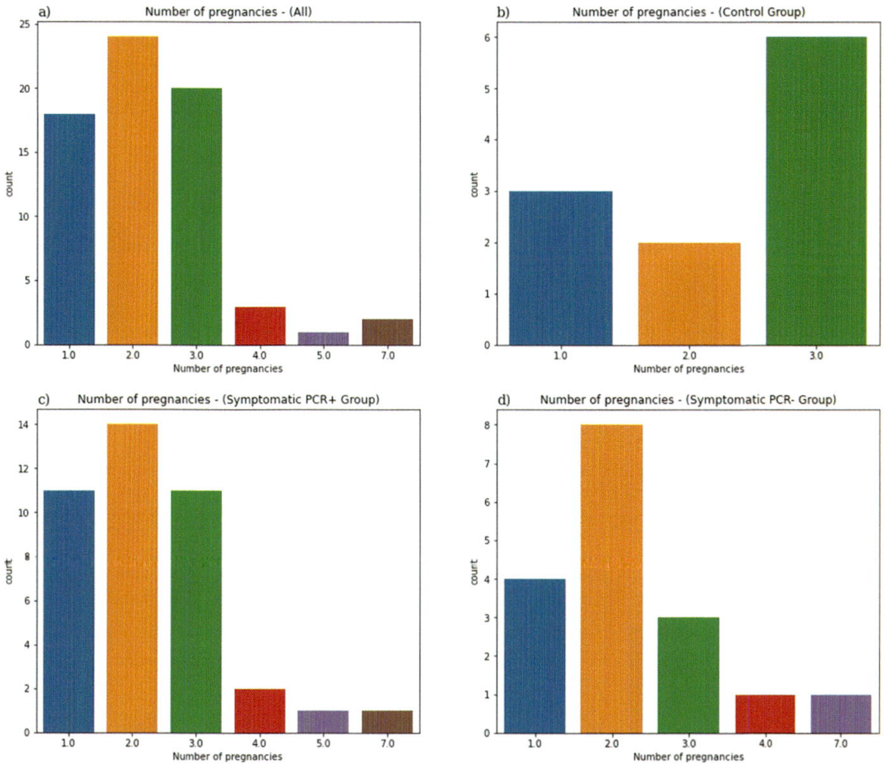

Fig. 7 Group characterization—number of pregnancies. Frequency distribution analysis considering **a** All the participants, **b** Control group, **c** Symptomatic PCR+ group, and **d** Symptomatic PCR- group

Considering the number of prior deliveries in Fig. 8, the highest frequency for the group with all participants was 1, followed by 2 and 0 (pregnant women in the first pregnancy). Similar behavior is found for the Symptomatic PCR+ group. For the Symptomatic PCR- group, the discrepancy between the most frequent number of prior deliveries, 1, was higher and was followed by 0 and 2. Finally, the Control group presented two as the most frequent number of prior deliveries.

4.2 *Symptoms*

The most relevant COVID-19 symptoms were collected from each patient during the anamnesis and are represented as a bar plot in Fig. 9. Since this is a direct characteristic of symptomatic patients, only two groups are presented, Symptomatic PCR+ in subfigure (a) and Symptomatic PCR- in subfigure (b).

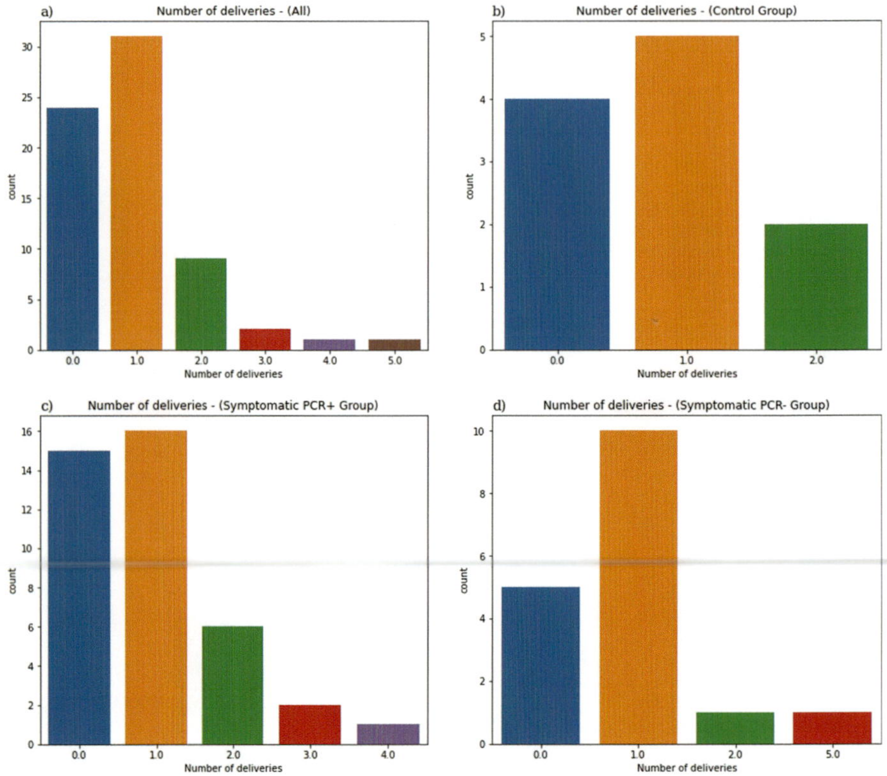

Fig. 8 Group characterization—number of prior deliveries. Frequency distribution analysis considering **a** All the participants, **b** Control group, **c** Symptomatic PCR+ group, and **d** Symptomatic PCR- group

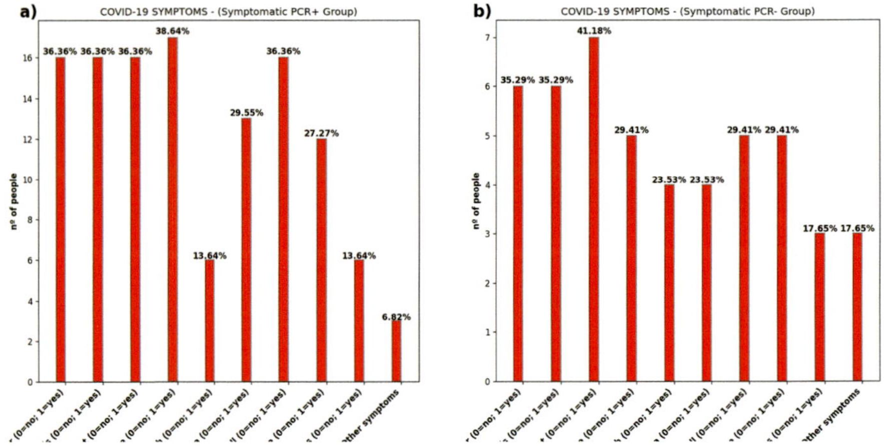

Fig. 9 Symptoms characterization for **a** Symptomatic PCR+ group, and **b** Symptomatic PCR- group

The most common symptom for the Symptomatic PCR+ group was Headache for 38.64% of the patients, followed closely by four other symptoms, Fever, Chills, Sore Throat, and Loss of Smell, with 36.36% each. Coryza and Loss of Taste were also present in almost 30% of the patients. Cough and Gastrointestinal Symptoms were present for only 13.64% of the patients each.

For the Symptomatic PCR- group, the most frequent symptom was Sore Throat for 41.18% of the patients, followed by Fever and Chills with 35.28% of the participants. Headache, Loss of Taste, and Loss of Smell were the third most cited symptoms, with 29.41% each. For this group, Cough and Coryza were also present for 23.53% of the patients, while Gastrointestinal and other Symptoms were present for 17.65% of the patients.

An explanation for the presence of symptoms with or without PCR+ may be that the patients' were in different phases of the disease. PCR+ patients would be in an earlier phase with viral detection and PCR- would be in an intermediary/inflammatory phase of the disease, without viral detection.

4.3 Maternal Health Status

This analysis presents a frequency distribution of maternal health status considering the occurrence of two relevant issues: pre-eclampsia and gestational diabetes.

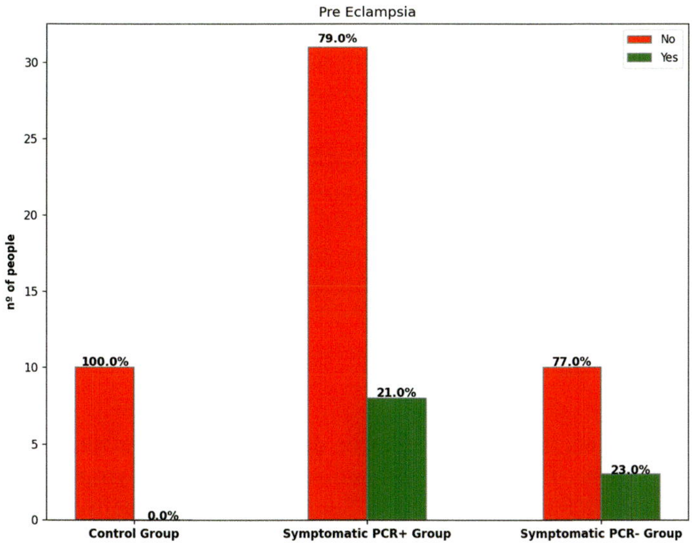

Fig. 10 Maternal Health Status—occurrence of gestational diabetes for the Control, Symptomatic PCR+, and Symptomatic PCR- Groups

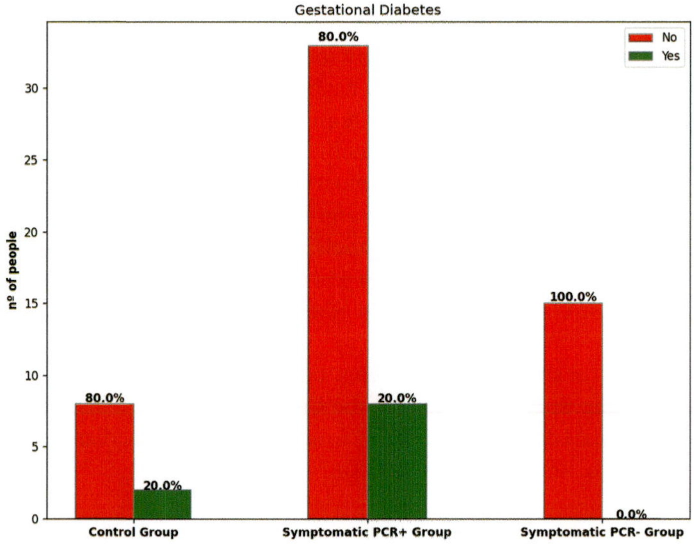

Fig. 11 Maternal Health Status—occurrence of Pre-eclampsia for the Control, Symptomatic PCR+, and Symptomatic PCR- Groups

In Fig. 10, the Symptomatic PCR+ presented 21% of cases of pre-eclampsia, while the Symptomatic PCR- group presented 23% of the patients. For the Control group, no cases of the condition were registered. This result indicates that the presence of COVID-19 symptoms (despite the PCR result) might increase the risk of pre-eclampsia. As previously discussed in this Chapter, pregnant women infected with COVID-19 are already a well-documented risky group, and this information reinforces the fact.

For the cases of gestational diabetes, presented in Fig. 11, 20% of the patients from both Symptomatic PCR+ and Control groups developed the disease. In contrast, the Symptomatic PCR- group has not registered any cases. In this case, no correlation with the presence of COVID-19 symptoms can be determined.

4.4 Birth Parameters

This Section presents the analysis of three different aspects related to birth. First, the dispersion characterization of the gestational age when the symptoms started and when the delivery effectively happened. Second, the type of delivery for the three groups is analyzed.

In Fig. 12, the boxplots of the gestational age at the start of the symptoms are presented for the Symptomatic PCR+ and Symptomatic PCR- groups. The analysis

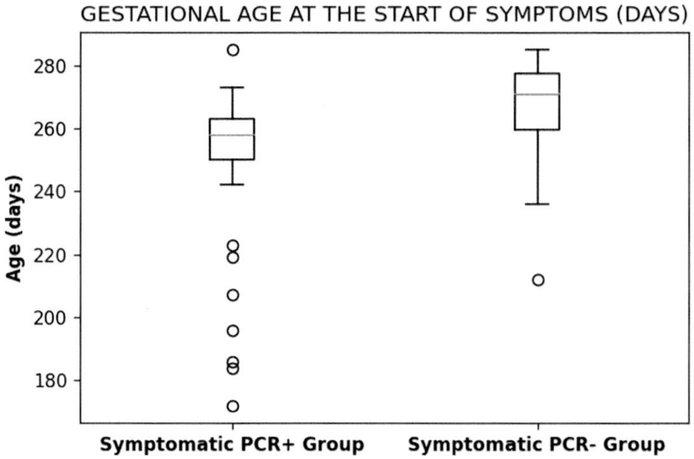

Fig. 12 Birth parameters—gestational age at the start of the symptoms. Only two groups presented COVID-19 symptoms

aims to characterize the groups since the start of the symptoms is not a controlled variable.

In contrast, in Fig. 13, the boxplots of the gestational age at birth are presented for the Control, Symptomatic PCR+, and Symptomatic PCR- groups. In this case, it is possible to provide a more detailed interpretation. Analyzing the groups through visual inspection, the Control and Symptomatic PCR- groups presented a median gestational age at birth higher than the Symptomatic PCR+ group. This is a consequence of the protocol followed by the MEAC Maternity Hospital, given the high risk of worsened COVID-19 symptoms and maternal death. When a pregnant woman was diagnosed with COVID-19 (usually confirmed with positive PCR) during the third trimester of pregnancy, immediate delivery, if possible, was prescribed.

Finally, the type of delivery is presented in Fig. 14 for the Control, Symptomatic PCR+, and Symptomatic PCR- groups. For this aspect, the protocols followed by the Maternity Hospital were followed.

Following the normal protocol, 75% of the deliveries were vaginal for the Control group, while 25% were cesarean sections. Similar statistics were found for both Symptomatic groups (PCR+ and PCR-). Following the COVID-19 protocol, 94% of the deliveries were cesarean, 3% cesarean with forceps, and 3% were vaginal deliveries. It is essential to highlight that during the pandemic, the decision for cesarean section for patients with COVID-19 symptoms and term pregnancy was to prevent the worsening of the infection.

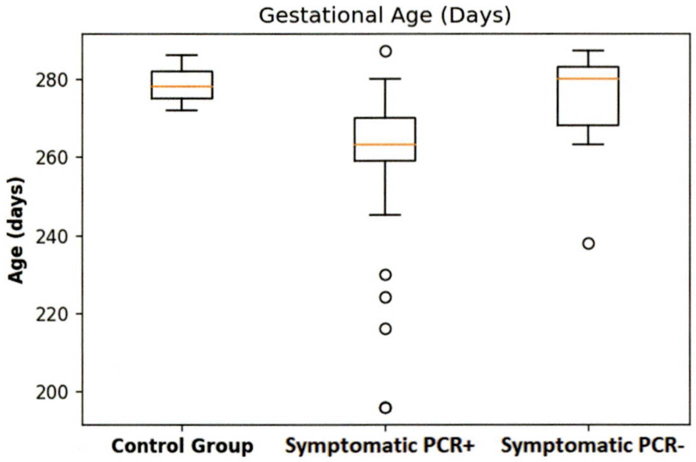

Fig. 13 Birth parameters—gestational age at birth for the three groups under study

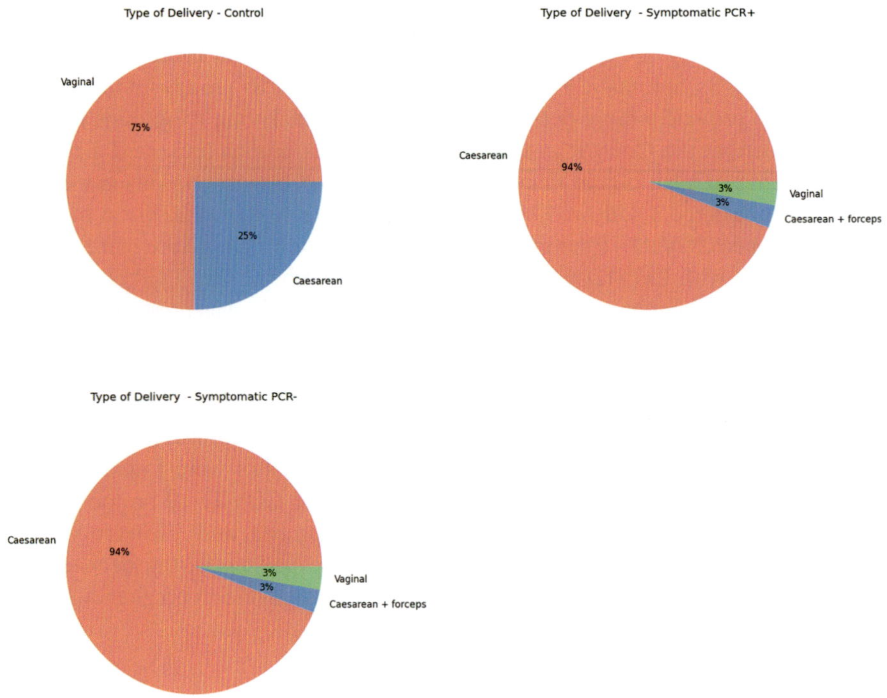

Fig. 14 Birth parameters—type of Delivery. **a** Control group, **b** Symptomatic PCR+ group, and **c** Symptomatic PCR- group

4.5 COPE Survey—Maternal Stress and Mental Health

As previously explained, the participants were submitted to the Portuguese version of the COVID-19 in Pregnancy and Early Childhood (COPE) survey. This Section covers a subset of questions related to Stress and Mental Health conditions.

The frequency distribution analysis of Question 27 of the COPE survey, "How has the COVID-19 outbreak changed your stress levels or mental health?" is presented in Fig. 15.

For this question, all the groups stated in their majority that their stress or mental health "moderately worsened" due to the COVID-19 outbreak. At the same time, no participants declared that they "moderately improved" or "significantly improved" in this aspect. The "no change" and "significantly worsened" were at similar levels

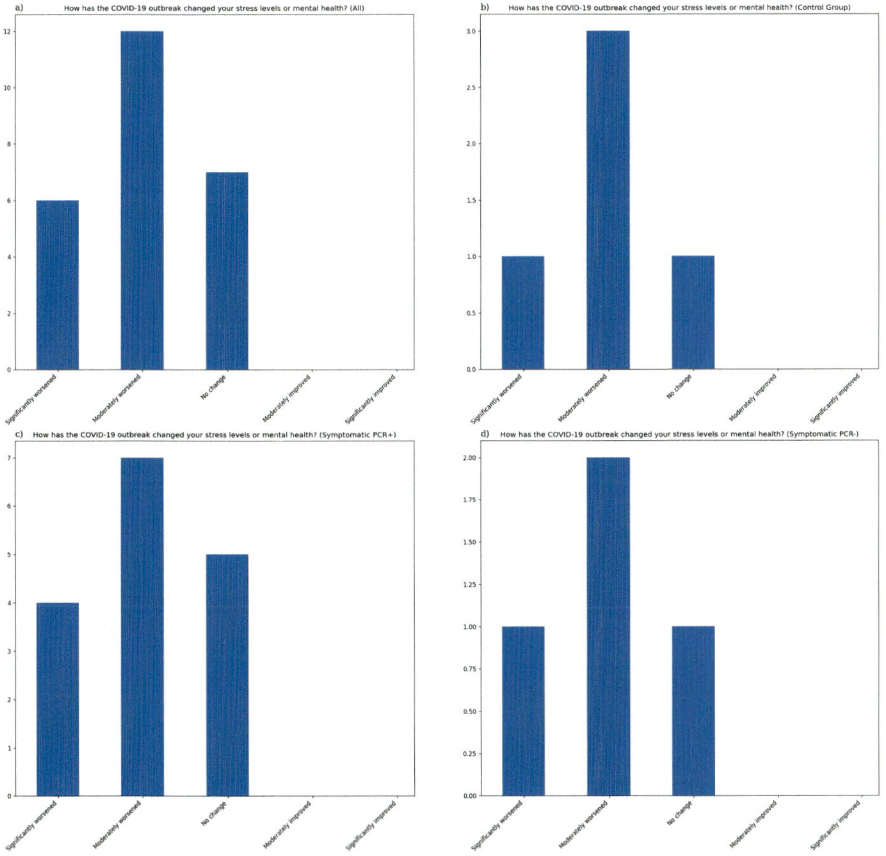

Fig. 15 COPE Survey Question 27: "How has the COVID-19 outbreak changed your stress levels or mental health?". Frequency distribution analysis considering **a** All the participants, **b** Control group, **c** Symptomatic PCR+ group, and **d** Symptomatic PCR- group

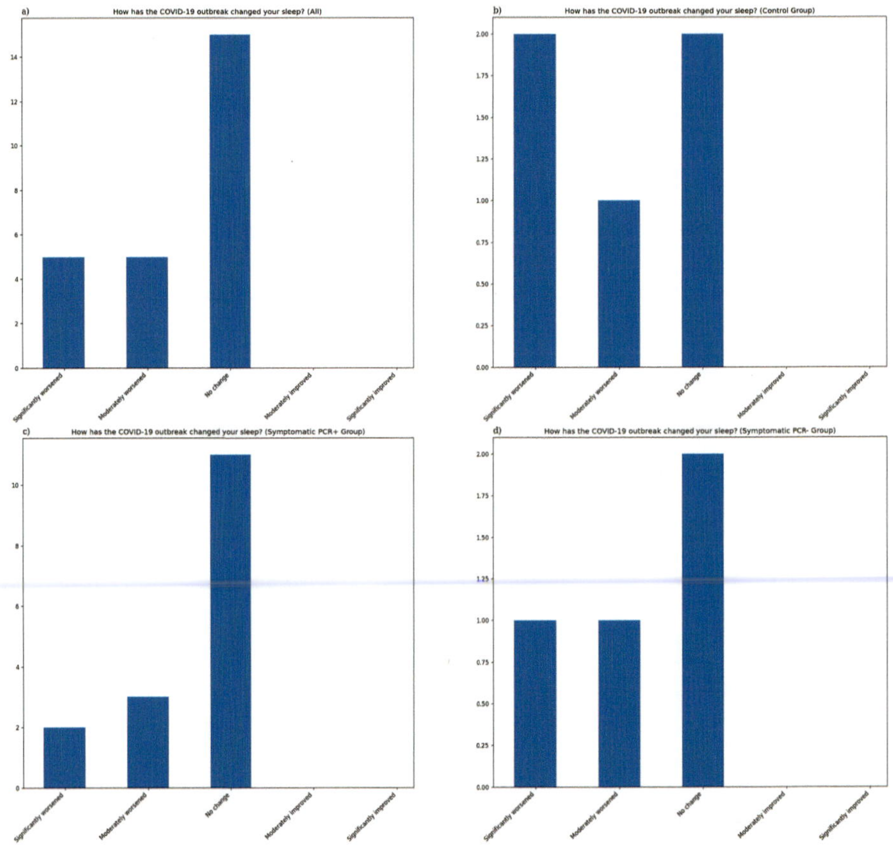

Fig. 16 COPE Survey Question 28: "How has the COVID-19 outbreak changed your sleep?". Frequency distribution analysis considering **a** All the participants, **b** Control group, **c** Symptomatic PCR+ group, and **d** Symptomatic PCR- group

for the three groups. Considering the two categories related to worsened conditions, it is possible to measure the significant negative impact of the COVID-19 outbreak on the stress and mental health levels of the participants.

The frequency distribution analysis of Question 28 of the COPE-IS survey, "How has the COVID-19 outbreak changed your sleep?" is presented in Fig. 16.

For Question 28, the majority of the participants of Symptomatic patients (PCR+ and PCR-) stated that their sleep levels haven't changed due to the pandemic. The Control group declared similar "no change" and "significantly worsened" levels. Again, no participant declared that they moderately or significantly improved their sleeping time. Let's consider the two categories related to worsened conditions. It is possible to measure the significant negative impact of the COVID-19 outbreak on the sleeping time of the participants, especially for the Control and Symptomatic PCR- groups.

The frequency distribution analysis of Question 29 of the COPE-IS survey, "How has the COVID-19 outbreak changed your daily energy levels?" is presented in Fig. 17.

For this question, most of the participants of Symptomatic PCR+ patients stated that their daily energy levels haven't changed due to the pandemic. The Control and Symptomatic PCR groups declared similar levels for "no change" and "moderately worsened". No participant from Control and Symptomatic PCR- groups declared that they moderately or significantly improved their daily energy levels, while a small number of participants in the Symptomatic PCR+ group declared that they moderately improved their energy levels. Let's consider the two categories related to worsened conditions. It is possible to indicate a neutral, negative impact of the COVID-19 outbreak on the participants' energy levels, but still with a significant

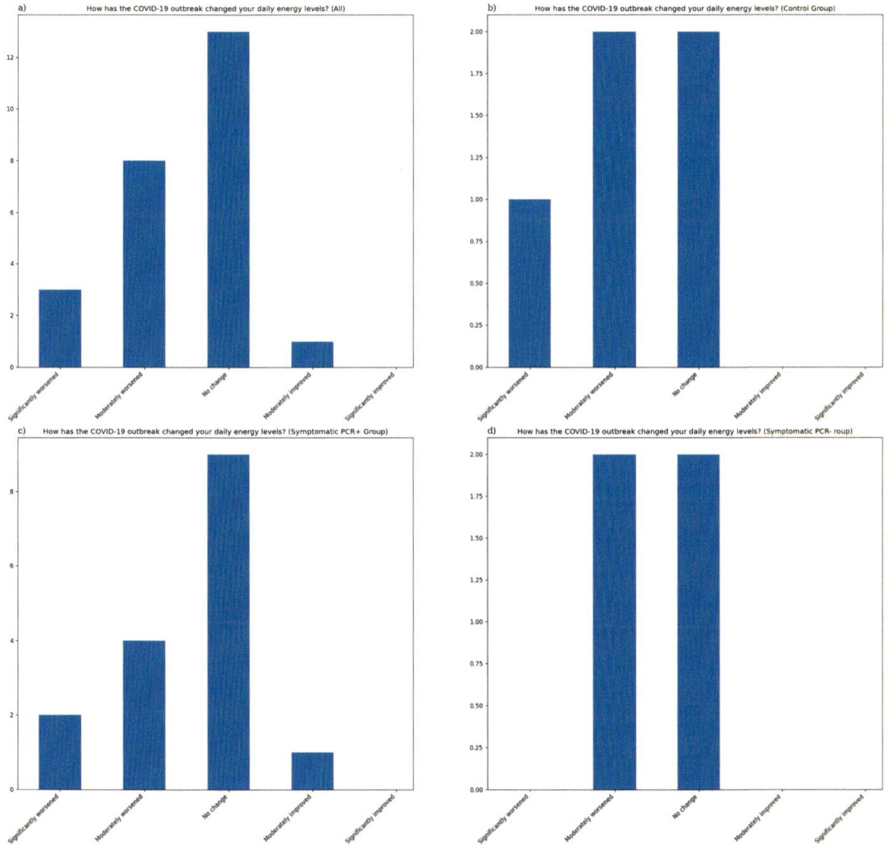

Fig. 17 COPE Survey Question 29: "How has the COVID-19 outbreak changed your daily energy levels?". **a** Frequency distribution considering all the participants, **b** Control group, **c** Symptomatic PCR+ group, and **d** Symptomatic PCR- group

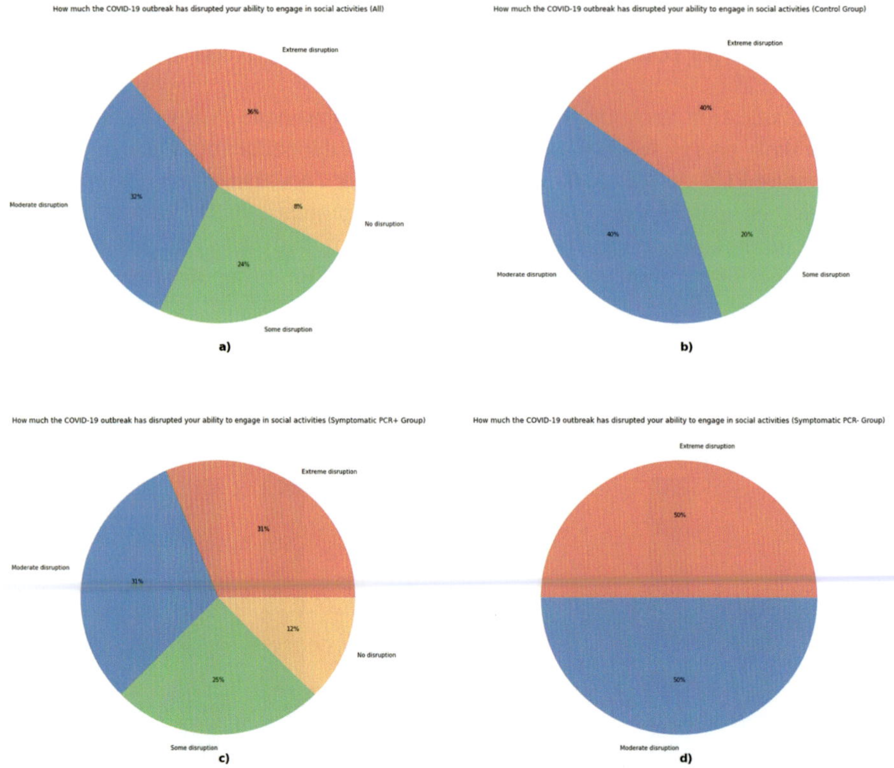

Fig. 18 COPE-IS Survey Question 30, sub-item a: "How much the COVID-19 outbreak has disrupted your ability to engage in social activities?" Frequency distribution analysis considering **a** All the participants, **b** Control group, **c** Symptomatic PCR+ group, and **d** Symptomatic PCR- group

number of participants declared that the energy levels moderately or significantly worsened for the three groups.

The frequency distribution analysis of Question 30, sub-item a, of the COPE-IS survey, "How much the COVID-19 outbreak has disrupted your ability to engage in social activities?" is presented in Fig. 18.

Analyzing the four plots, it is possible to find a substantial variation in the participants' answers. This might be because of the perception of what kind of social activities they considered when answering the survey. The "No Disruption" percentages were 20% for the Control group, 31% for the Symptomatic PCR+ group, and 50% for the Symptomatic PCR- group. With that, a vast majority of participants declared that the pandemic caused "Some", "Moderate", and "Extreme" disruption in their ability to engage in social activities.

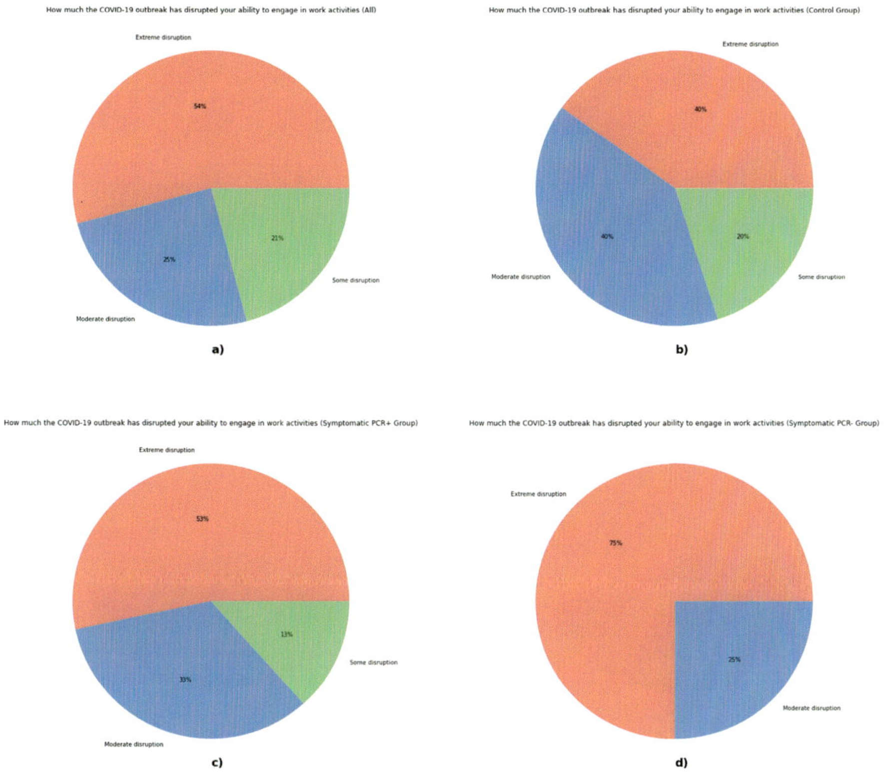

Fig. 19 COPE-IS Survey Question 30, sub-item b: "How much the COVID-19 outbreak has disrupted your ability to engage in work activities?". Frequency distribution analysis considering **a** All the participants, **b** Control group, **c** Symptomatic PCR+ group, and **d** Symptomatic PCR- group

The frequency distribution analysis of Question 30, sub-item b, of the COPE-IS survey, "How much the COVID-19 outbreak has disrupted your ability to engage in work activities?" is presented in Fig. 19.

Analyzing the Figure, it is possible to find a significant share of "no disruption" for the Symptomatic PCR+ and Symptomatic PCR- groups, with 53% and 75%, respectively. On the opposite direction, the "Extreme Disruption" shares were 20% for the Control group, 33% for the Symptomatic PCR+ group, and no responses for the Symptomatic PCR- group.

The frequency distribution analysis of Question 30, sub-item c, of the COPE-IS survey, "How much the COVID-19 outbreak has disrupted your ability to engage in physical activities?" is presented in Fig. 20.

In the case of physical activities, the negative impact presented in the graphical analysis is substantial. Firstly, analyzing all the participants' answers, only 8% declared "No change" and "Moderately improved", with 40% of "Moderately worsened" and 44% "Significantly worsened". For the Symptomatic PCR- group, half

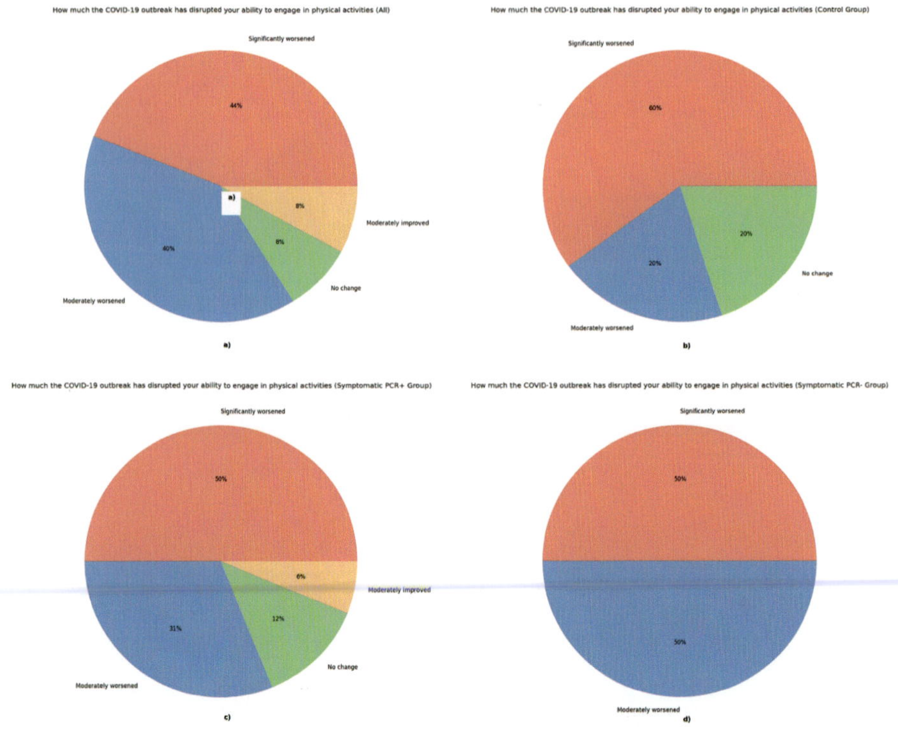

Fig. 20 COPE-IS Survey Question 30, sub-item c: "How much the COVID-19 outbreak has disrupted your ability to engage in physical activities?". Frequency distribution analysis considering **a** All the participants, **b** Control group, **c** Symptomatic PCR+ group, and **d** Symptomatic PCR- group

of the patients answered "Moderately worsened" and the second half, "Significantly worsened". "Significantly worsened" was also the answer of 60% of the Control group and 50% of the Symptomatic PCR+ group.

The frequency distribution analysis of Question 30, sub-item d, of the COPE-IS survey, "How much the COVID-19 outbreak has disrupted your access to healthy meals?" is presented in Fig. 21.

The possibility of having access to healthy meals was also strongly negatively impacted during the COVID-19 outbreak. The data analysis shows that 94% of the patients faced disruption in this aspect. No participants from the Control group answered "No disruption", with 50% facing "Extreme disruption". In the case of Symptomatic PCR+ patients, 69% faced "Extreme disruption", while 50% declared the same answer in the Symptomatic PCR- group.

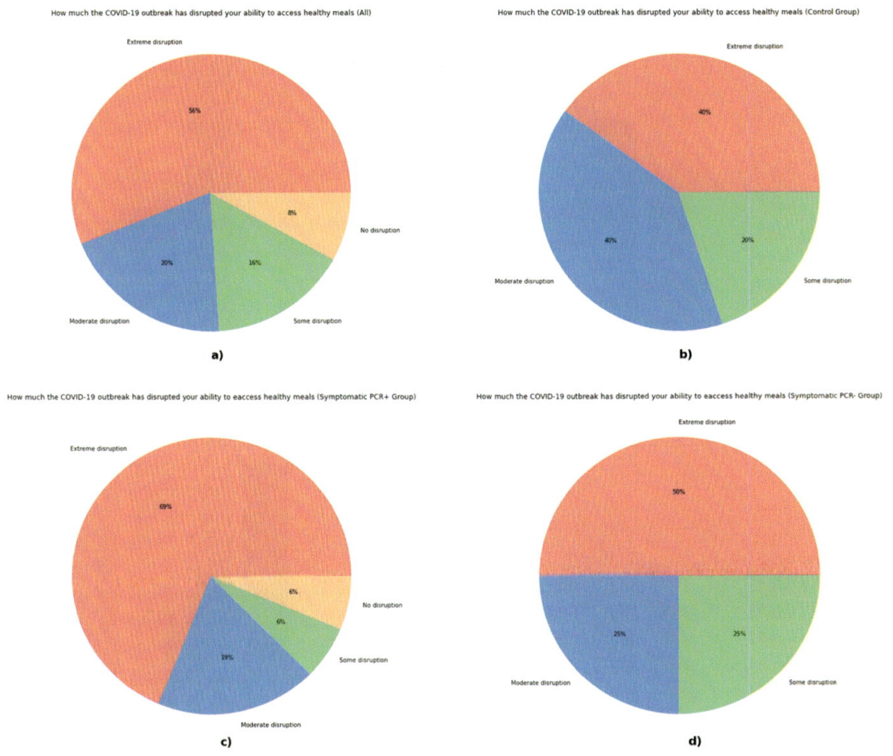

Fig. 21 COPE-IS Survey Question 30, sub-item d: "How much the COVID-19 outbreak has disrupted your access to healthy meals?". Frequency distribution analysis considering **a** All the participants, **b** Control group, **c** Symptomatic PCR+ group, and **d** Symptomatic PCR- group

The frequency distribution analysis of Question 31 of the COPE-IS survey, "Overall level of impact on your daily life as a result of the COVID-19 outbreak", is presented in Fig. 22. This question is a Likert-scale type, where 1 indicates the lowest impact, 4 indicates neutral impact, and 7 indicates the maximum impact.

The graphics clearly indicate a significantly high level of impact on the patient's daily lives, with a higher frequency of answers starting on the 4 and going up. No participant from the Symptomatic PCR+ group answered less than 3.

The frequency distribution analysis of Question 32 of the COPE-IS survey, "General level of stress related to the COVID-19 outbreak", is presented in Fig. 23. This question is a Likert-scale type, where 1 indicates the lowest level of stress, and 7 indicates the maximum level of stress.

In this case, the participant's answers were more dispersed, but still with the highest number of answers in the highest level of stress when considering all the participants. The Control group was the one declaring lower levels of stress. There is certainly an impact caused by the conditions of these patients, pregnant women, especially the ones presenting COVID-19 symptoms.

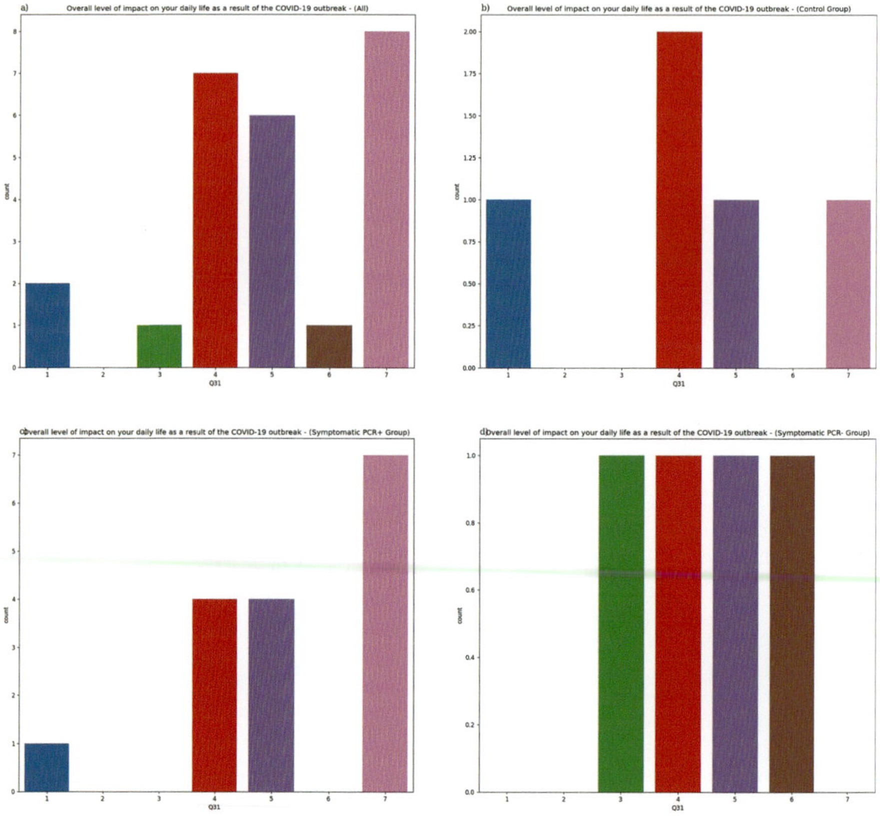

Fig. 22 COPE-IS Survey Question 31: "Overall level of impact on your daily life as a result of the COVID-19 outbreak". Frequency distribution analysis considering **a** All the participants, **b** Control group, **c** Symptomatic PCR+ group, and **d** Symptomatic PCR- group

The frequency distribution analysis of Question 33 of the COPE survey, "Extent to which you see the COVID-19 outbreak as having a positive or negative impact on your life", is presented in Fig. 24. This question is a Likert scale type, where 1 indicates the highest negative impact, 4 indicates neutral impact, and 7 indicates the highest positive impact.

This question presents an inverse perception if compared to the other Likert-scale previously presented. In this case, the highest the answer, the more positive the perception. Analyzing the graphical representation, the greater majority of the responses are concentrated on levels 1 and 2 for the three groups under analysis and also for all the participants.

The frequency distribution analysis of Question 34 of the COPE-IS survey, "How long do you think it will take before things get back to normal?" is presented in Fig. 25.

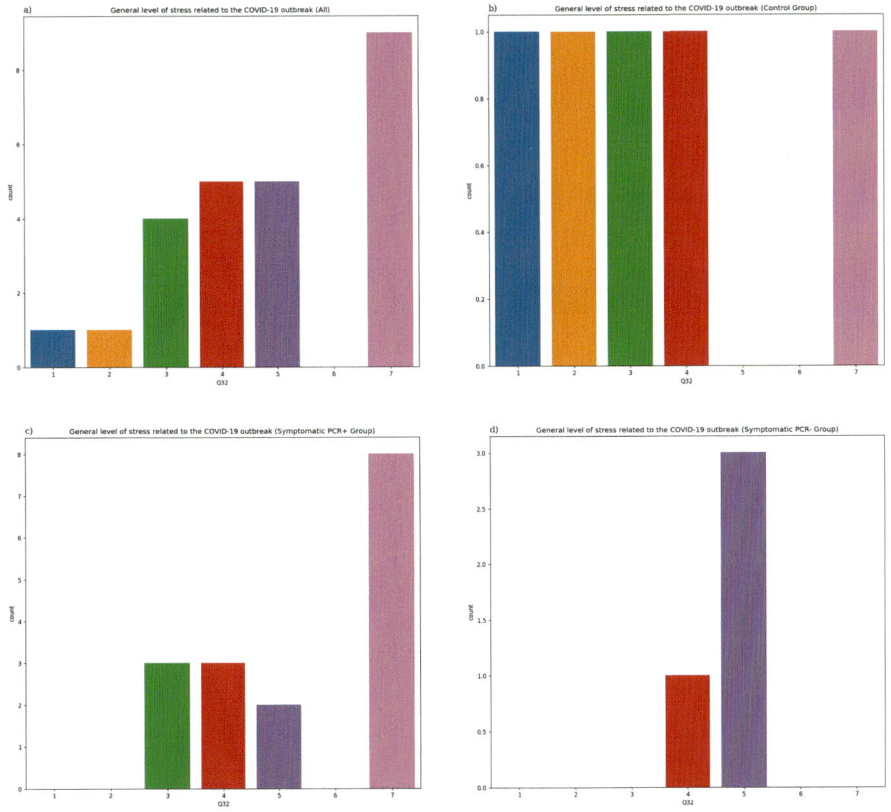

Fig. 23 COPE-IS Survey Question 32: "General level of stress related to the COVID-19 outbreak". Frequency distribution analysis considering **a** All the participants, **b** Control group, **c** Symptomatic PCR+ group, and **d** Symptomatic PCR- group

The results indicate that, even with facing severe negative impacts from the pandemic in their daily lives, such as extreme disruption in having access to healthy meals and performing physical activities, and higher levels of stress, 72% of all the participants stated that the normality would be restored in less than one month. The pattern was the same for the three groups under analysis, with a total of 60%, 69%, and 100% of the Control, Symptomatic PCR+, and Symptomatic PCR-, respectively.

The frequency distribution analysis of Question 35 of the COPE-IS survey, "What is your most significant source of stress due to the COVID-19 outbreak right now?" is presented in Fig. 26. This question allowed many possible answers, and the graphics present only the items with answers for each group of participants.

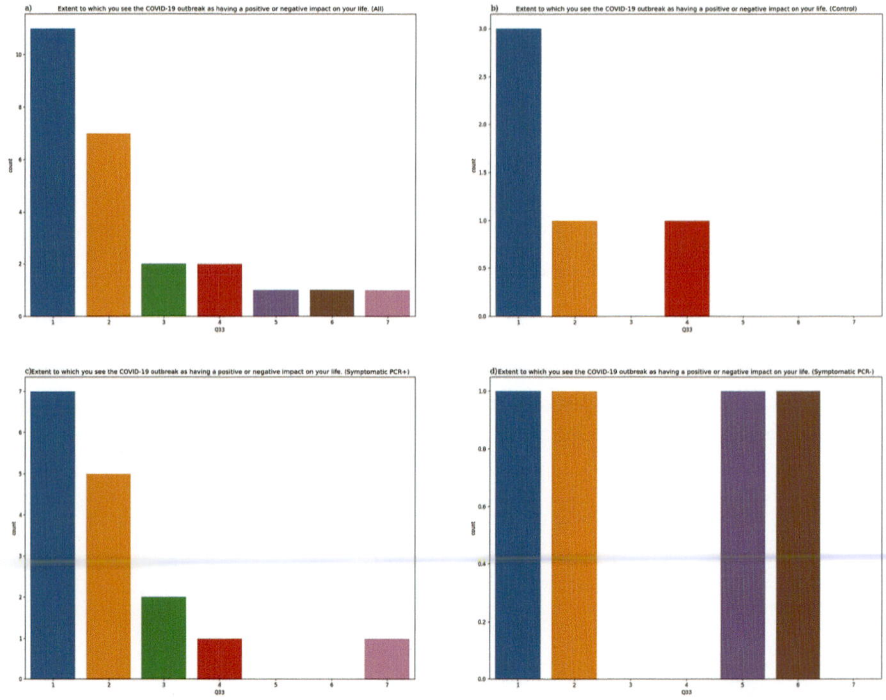

Fig. 24 COPE-IS Survey Question 33: "Extent to which you see the COVID-19 outbreak as having a positive or negative impact on your life". Frequency distribution analysis considering **a** All the participants, **b** Control group, **c** Symptomatic PCR+ group, and **d** Symptomatic PCR- group

Personal constraints strongly influence sources of stress, and seven items were selected. Considering all the participants, "Health concerns" was the primary source for 32% of the respondent, with "Financial concerns" with 24% as the second. Impact on others (baby, partner, community, family members, and close friends) was also indicated as a source of stress. The Control and the Symptomatic PCR- only declared "health" and "financial" concerns together with "Impact on your baby".

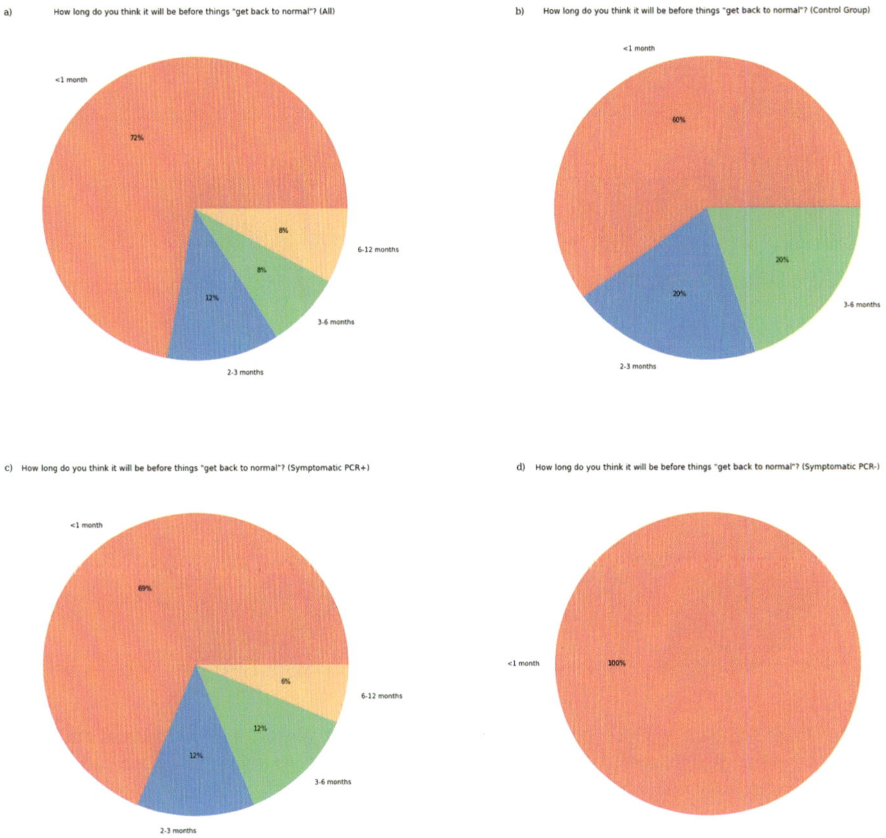

Fig. 25 COPE-IS Survey Question 34: "How long do you think it will take before things get back to normal?". Frequency distribution analysis considering **a** All the participants, **b** Control group, **c** Symptomatic PCR+ group, and **d** Symptomatic PCR- group

5 Conclusion

This chapter comprehensively characterized multiple clinical and survey data on the emotions and feelings of pregnant women from the public Maternity Hospital of the Federal University of Ceará (MEAC) in Fortaleza, Brazil. An Exploratory Data Analysis (EDA) using graphical and statistical techniques was performed.

The scientific literature presents strong indications that pregnant women with COVID-19 are at an increased risk for developing more severe illness conditions, including admission to the ICU and a higher probability of death. In addition, multiple risk factors are mapped, such as infection during the third trimester of pregnancy, previous health issues, and being overweight, among others.

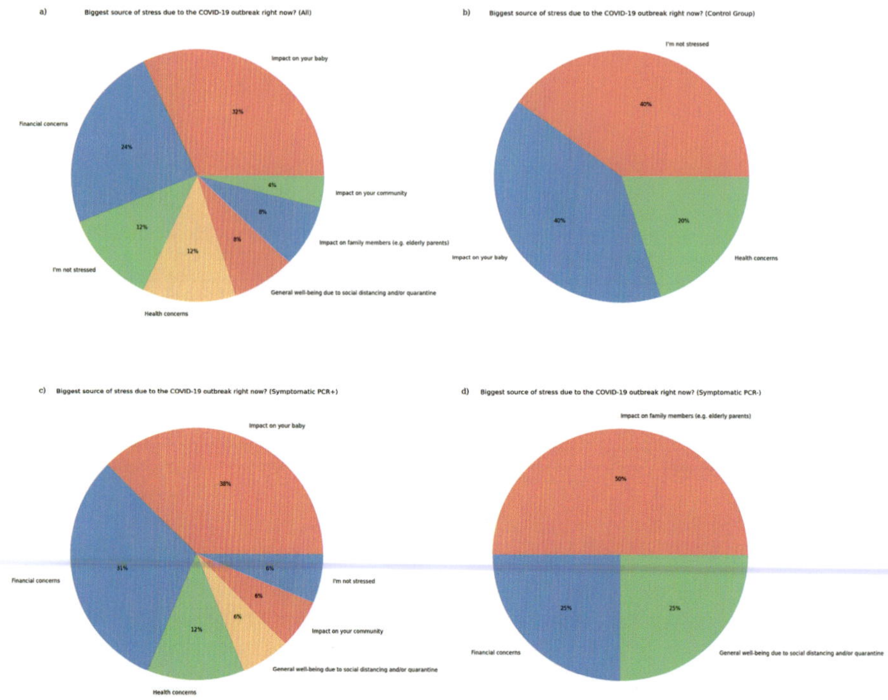

Fig. 26 COPE-IS Survey Question 35: "What is your most significant source of stress due to the COVID-19 outbreak right now?". Frequency distribution analysis considering **a** All the participants, **b** Control group, **c** Symptomatic PCR+ group, and **d** Symptomatic PCR- group

The analysis provided here aims to contribute to a better understanding of the mechanisms related to the impact of SARS-CoV-2 infection on maternal health status and birth parameters, as well as to investigate the impact of the COVID-19 outbreak on stress levels and mental health of the groups under study, which may give support to effective interventions in the future.

Acknowledgements The authors would like to thank:
The Brazilian National Council for Scientific and Technological Development (CNPq).
Staff from the Assis Chateaubriand Maternity Public Hospital at the Federal University of Ceará (MEAC-UFC).
The research team from the Laboratory of Applied Neurosciences at the University of Saint Joseph, Macau SAR, China.

References

1. Sheffield J (2022) Coronavirus and pregnancy: what you should know. https://www.hopkinsmedicine.org/health/conditions-and-diseases/coronavirus/coronavirus-and-covid-19-what-pregnant-women-need-to-know. Accessed 04 Jan 2023
2. NHS Response Team (2022) Pregnancy and coronavirus (covid-19). https://www.nhs.uk/pregnancy/keeping-well/pregnancy-and-coronavirus/. Accessed 04 Jan 2023
3. COVGEN (2023) Covgen research alliance. https://www.covgen.org/. Accessed 24 Jan 2023
4. Zimbron J, Dalman C, Lewis G, Jones PB, Khandaker GM (2012) Childhood infection and adult schizophrenia: a meta-analysis of population-based studies. Schizophr Res 139(1–3):161–168
5. Iyegbe CO, Vernon AC, Yolken R, Murray RM, Pollak TA, Kępińska AP (2020) Schizophrenia and influenza at the centenary of the 1918-1919 Spanish influenza pandemic: mechanisms of psychosis risk. Front Psych 11(1)
6. Rasmussen SA, Jamieson DJ (2022) An update on covid-19 and pregnancy. Amer J Obstet Gynecol 226(2):177–186
7. Liu S, Wei SQ, Bilodeau-Bertrand M, Auger N (2021) The impact of covid-19 on pregnancy outcomes: a systematic review and meta-analysis. Canad Med Assoc J - CMAJ 193(61):E540–E548
8. Shaman JL et al, Emeruwa UN, Ona S (2020) Associations between built environment, neighborhood socioeconomic status, and sarscov-2 infection among pregnant women in New York city. J Amer Med Assoc 2(1):324–390
9. Badell ML, Horton JP, Boulet SL, Jamieson DJ, Joseph NT, Stanhope KK (2020) Sociodemographic predictors of sars-cov-2 infection in obstetric patients. Emerg Infect Dis 26(1):2787–9
10. Li J, Zhang T, Wang D, Lan W, Liu H, Liu F (2020) Clinical and ct imaging features of the covid-19 pneumonia: focus on pregnant women and children. J Infect 80(5):e7–e13
11. Tang F, Peng M, Gao Y, Peng J, Xie H, Zhao Y, Jin Z, Sun G (2020) Clinical features and outcomes of pregnant women suspected of coronavirus disease 2019. J Infect 81(1):e40–e44
12. Erol SA, Yucel Yetiskin FD, Besimoglu B, Ozden Tokalioglu E, Anuk AT, Turgut E, Goncu Ayhan S, Turgay B, Unlu S, Kanmaz G, Dinc B, Ozgu-Erdinc AS, Keskin HL, Surel AA, Tekin OM (2022) Management of pregnant women with covid-19: a tertiary pandemic center experience on 1416 cases. J Med Virol 94(3):1074–1084
13. Skouteris H, Byrne LK, Hayden MJ, Davies SJ, Lum JAG (2018) Cognitive impairment during pregnancy: a meta-analysis. Med J Austr 208(1):35–40

GPSR Compliance

The European Union's (EU) General Product Safety Regulation (GPSR) is a set of rules that requires consumer products to be safe and our obligations to ensure this.

If you have any concerns about our products, you can contact us on ProductSafety@springernature.com

In case Publisher is established outside the EU, the EU authorized representative is:

Springer Nature Customer Service Center GmbH
Europaplatz 3
69115 Heidelberg, Germany

Batch number: 10756376

Printed by Printforce, the Netherlands